建筑立场系列丛书 No.33

本土现代化
Vernacular and Modern

中文版

韩国C3出版公社 | 编

李海玲 高翔 杨开红 于风军 | 译

大连理工大学出版社

4 本土现代化

- 004 本土现代化？_Jaap Dawson
- 012 圣乔治环境展览中心_Ana Laura Vasconcelos
- 028 托肯湖游客中心_Wingårdh Arkitektkontor AB
- 040 黛莱会议中心_Vo Trong Nghia + Takashi Niwa
- 050 曼谷树屋_Nuntapong Yindeekhun + Bunphot Wasukree
- 060 贝斯凉亭_H&P Architects
- 068 兰溪庭_Archi-Union Architects
- 074 肉桂培训中心_TYIN tegnestue Architects

084 乡土情怀

- 084 乡土情怀_Maurizio Scarciglia
- 088 9水疗中心_a21 Studio
- 098 意大利岩屋的Basiliani酒店_Domenico Fiore
- 108 Hornitos酒店_Gonzalo Mardones Viviani
- 122 Fasano Boa Vista酒店_Isay Weinfeld

136 金孝晚

- 136 传统景观_HyoMan Kim
- 138 嵌入的非线性场景的冲击效应_TaeCheol Kim + HyoMan Kim
- 142 形式追随功能?功能追随形式！_TaeCheol Kim
- 148 GaOnJai
- 164 KyeongDokJai
- 178 办公园

- 188 建筑师索引

Vernacular and Modern

004 *Modern Vernacular?* _ Jaap Dawson

012 *São Jorge Interpretation Center* _ Ana Laura Vasconcelos

028 *Tåkern Visitor Center* _ Wingårdh Arkitektkontor AB

040 *Dailai Conference Hall* _ Vo Trong Nghia + Takashi Niwa

050 *Bangkok Tree House* _ Nuntapong Yindeekhun + Bunphot Wasukree

060 *BES Pavilion* _ H&P Architects

068 *Lanxi Curtilage* _ Archi-Union Architects

074 *Cassia Coop Training Center* _ TYIN tegnestue Architects

VERNACULAR FICTION

084 *Vernacular Fiction* _ Maurizio Scarciglia

088 *9 Spa* _ a21 Studio

098 *Basiliani Hotel in the Rock Dwellings of Italy* _ Domenico Fiore

108 *Hornitos Hotel* _ Gonzalo Mardones Viviani

122 *Fasano Boa Vista Hotel* _ Isay Weinfeld

HyoMan Kim

136 *Tradition Scaping* _ HyoMan Kim

138 *The Shock Effect of Inserted Nonlinear Scenes* _ TaeCheol Kim + HyoMan Kim

142 *Form follows function? Function follows form!* _ TaeCheol Kim

148 GaOnJai

164 KyeongDokJai

178 Office Park

188 Index

本土现代化

如果我们是建筑师，我们就是现代建筑师。我们不但学习过建筑传统——组合、施工和细部处理。我们还接受过训练，使我们的设计属于现代，或者说不只属于一个特定的时代。

其他时代和带有其他文化特点的建筑师们和建设者能够分享现代建筑理念吗？他们能够找到一种有别于现代建筑理念的新思路或是新方法吗？我们能把现代建筑理念和那种永恒的、本土化的建筑理念相结合吗？

在近代建筑设计中，要做到这一点非常困难。本土化的建筑不只是建筑工艺、材料或是形式，而是一种态度——那是一种让我们从自己建造的事物中看到自身存在的态度。毫无疑问，我们在容纳身体的空间里认知自己，而不仅仅是为了一个"理念"。建筑通过建筑方法、规模和不同空间范围的结合来为我们的身体营造存在的空间。

如果我们是为了我们的身体营造存在的空间，我们就应该为我们的"整体存在"营造空间。我们的目标既不是证明我们是现代的或是创新的，而是要为我们的"整体存在"营造空间。

If we are architects, we are modern architects. We have been trained not only in the modern tradition of composition and construction and details. We have also been trained to regard our designs as modern or not, as belonging to a particular time.

Did architects and builders in other ages and other cultures share our modern attitude? Can architects and builders now discover an attitude and a way of building something different from the modern attitude? Can we combine modern insights with timeless, vernacular insights?

Recent designs reveal how difficult it is. Vernacular building, it turns out, is more than techniques and materials and forms. Vernacular building is an attitude. It is an attitude that lets us see ourselves in what we build. Without thinking about it, we recognize ourselves in spaces that make space for our body, not alone space for our concepts. They make space for our body through their measures, their scales, and the composition of their boundaries.

If we build in order to make space for our body, then we make space for our whole being. Our goal is not to prove we're contemporary, not to show we're innovative. Our goal is to make space for our whole being.

Vernacular and Modern

圣乔治环境展览中心/Ana Laura Vasconcelos
托肯湖游客中心/Wingårdh Arkitektkontor AB
黛莱会议中心/Vo Trong Nghia + Takashi Niwa
曼谷树屋/Nuntapong Yindeekhun + Bunphot Wasukree
贝斯凉亭/H & P Architects
兰溪庭/Archi-Union Architects
肉桂培训中心/TYIN tegnestue Architects

本土现代化?/Jaap Dawson

São Jorge Interpretation Center/Ana Laura Vasconcelos
Tåkern Visitor Center/Wingårdh Arkitektkontor AB
Dailai Conference Hall/Vo Trong Nghia + Takashi Niwa
Bangkok Tree House/Nuntapong Yindeekhun + Bunphot Wasukree
BES Pavilion/H & P Architects
Lanxi Curtilage/Archi-Union Architects
Cassia Coop Training Center/TYIN tegnestue Architects

Modern Vernacular?/Jaap Dawson

本土现代化?

本土现代化矛盾的。这两个词意是冲突的、意义相反的。它们没有交集。就如你说一件物体是巨大的渺小，我们会哑然失笑，可是当我们寻找本土现代化的建筑时，我们可能意识不到其中自相矛盾的地方。

本土化就是本地化。本土化是源于语言研究，只在最近才用于建筑。本地化来自我们的生活经历。在建筑业，我们对来自本地的建筑材料和工艺和从遥远的外地运来的材料和外来的建筑工艺是有区分的。

我们不只是和物质世界的材料打交道，我们也在探索这个世界的意义。当我们看到我们称之为本土化的建筑时，我们在这些建筑里看到了自己。当我们穿堂入室时，竖立的窗户就像人一样站在那里，窗户不像邮箱那样会倒下，整体建筑就像一个人形，立于地面，头戴王冠。

它们不是抽象的雕塑、空洞的事物。立柱也不会毫无意义地把力量插入地面，它们像有生命一样，有脚、身体和头，向我们致敬。本土化建筑中房间的大小也恰好能满足我们的身体和心灵的原本需要。房间也自然地具备本土化特点，围绕着中心的空间，聚为一体。

本土化的特点，是我们本地的特质，是在我们身上自然形成的。我们趋向于建造一个世界。在这个世界里，我们是生机勃勃的。为了建造这样的世界，我们唯一要做的就是做类比。在一座本土建筑中，我们所看见的、所应用的不仅仅是材料，还有类似于人类，类似于生命的元素。我们建造一座赋予我们意义的建筑，而不仅仅是为了舒适。我们建造设有安置灵魂的空间的世界，而不仅仅是为了容纳我们的活动。

Modern Vernacular?

Modern vernacular is an oxymoron. The two terms fight with each other. They are opposites. They don't belong together. If we say something is gigantically tiny, we smile at the oxymoron of the two terms form. But when we look for examples of modern vernacular buildings, we're probably not aware of the paradox.

Vernacular

Vernacular is what is native. We applied the term originally to language, and only later on to buildings. What's native is what arises in our experience. In building we distinguish native materials and building techniques from materials and ways of building we import from far away.

Our experience is not only our encounter with materials in a materialistic world. Our experience is also our discovery of meaning in our world. When we look at buildings we call vernacular, we see ourselves reflected in them. Windows stand, just like people stand when they move through rooms. Windows don't lie down, don't look like mailboxes. Elevations remind us of the human form: they stand on the ground, and they wear a crown. They are not abstract sculptures, disembodied things. Columns don't only serve to conduct annoying forces into the ground: they greet us as living beings with feet, body, and head. And the size of the rooms in vernacular buildings grows from our original need to contain our body and our soul. Vernacular rooms give way naturally to vernacular compositions: rooms grouped together around a central space.

Vernacular, is what's native to our own make-up, what arises in us automatically. It's our tendency to build a world we can experience as alive; and the only way to do that is through analogy. What we see, what we meet, in a vernacular building is not materials alone but elements that are like us, like living beings. We build a world that gives us meaning, not just comfort. We build a world that makes space for our soul, not just for our activities.

圣乔治环境展览中心采用本地材料建成,使其轻易地与周围环境融为一体

São Jorge Interpretation Center, made of vernacular materials so that it blends in easily with the surrounding environment

现代化

本土现代化是另一回事情。现代化属于一个特定的时期。我们知道没有比昨天的报纸更旧的东西了。它不是现代的东西,也没有了风格。

大约从20世纪20年代起,在建筑业,"现代化"的概念引起专业建筑师的重视。现代化变成了一个教条。它告诉我们应该无视过去的建造方式和我们在建筑业所秉承的理念——本土化的建筑理念。如果你仔细看一下我们定义为"现代化"的建筑,你就会发现一些明显的东西,这些"现代化"的建筑并不完全是全新的事物,它们只是对过去做出了否定。

它们否定过去的建筑,因为它们没有遵循组合的秩序,而这种秩序的制定源于我们对世界的认识。Christopher Alexander在反复研究了各个时期和各种文化的本土化建筑之后,精确地用15种特征描述了这种秩序。这些特征包括:在空间和材料的组合中,应突出和烘托中心、保持局部(不局限于局部)对称以及边界清晰等。Nikos Salingaros把这15个特征放在自然界中,举例说明了生物学和建筑设计中碎片的组织形式。

现代建筑遵循的规则和本土化建筑遵循的规则相悖。在现代化建筑中,我们没有发现局部对称,而是像机器一般的串联。我们找不到中心,也找不到清晰的边界。空间到处都是,又向四周流窜,空间不再容纳我们。整体的建筑立面不再是一个垂直的或者横向的层次,而是互不相连的抽象形状或者抽象形状的组合。也许对永恒建筑理念最大的否定就是现代建筑理念,即一座建筑看起来像是机器建造的。

建筑、空间、态度

让我们看看几个我们称之为本土现代化的建筑。我们来重新探寻一下在这些建筑中,哪些是本土化的元素,那些是现代化的元素。我们来分辨一下这些建筑的建筑师运用了哪些设计原则。一旦我们看到这些建筑,我们还要问一些其他问题——首先,为什么我们要给这

Modern

Modern is another kettle of fish altogether. Modern is what belongs to a particular time. And we all know there's nothing so old as yesterday's newspaper. It's no longer modern. It's no longer in style.

In architecture, and among professional architects since roughly the 1920s, modern is something else again. It's a doctrine. It tells us we should ignore both the ways we built previously and the attitude we previously had when we built: the vernacular attitude. If you take a good look at buildings we classify as modern, you will discover something quite revealing. They're not truly new. They're the denial of what we always built.

They're the denial of what we always built because they go out of their way and not to follow the compositional order that arises from how we perceive our world. Christopher Alexander adeptly describes this order in 15 properties he detects again and again in vernacular buildings through the ages and across cultures. The properties, in both the spaces and the materials that bound them, include focusing on and celebrating centres, local (but not rigid) symmetries, clear boundaries and etc.. Nikos Salingaros grounds the 15 properties in the natural world, showing for example the organizational pattern of fractals in biology and in design.

Modern architecture follows rules, and the design rules are the negation of the rules that lead to vernacular buildings. Instead of local symmetries, we find machine-like series. We can't find centres anymore, neither do we find clear boundaries. Space flows everywhere, escapes, no longer contains us. Rather than a vertical or horizontal hierarchy in elevations, we see abstract shapes or combinations of shapes that are not linked together. And perhaps the most glaring denial of timeless architecture is the modern doctrine that a building should look as though a machine had built it.

Buildings, Spaces, Attitudes

Let's look at several recent examples of buildings we might call modern vernacular. Let's rediscover what is vernacular in them and what is modern in them. Let's discern the design rules their architects followed. And once we've met the buildings, we'll be ready to ask other questions. Why are we concerned with labels in the first place? What do labels have to do with our experience of spaces and buildings? Why do we let a style straitjacket us? What's

托肯湖游客中心，通过升级传统的茅草屋与现代化的外形，来与历史进行对话

Tåkern Visitor Center, establishing a dialogue with history by updating the local tradition of thatched roofs with a modern form

黛莱会议中心，把人们带回自然，体验当地的生活

Dailai Conference Hall, leading people back to nature for a vernacular experience

些建筑插上标签呢？这些标签和我们对空间或是建筑的认识有什么联系呢？为什么我们要让一些风格约束我们呢？我们给建筑物插上"现代化"的标签，我们的动机是什么？

圣乔治环境展览中心

这座位于葡萄牙圣乔治岛的令人精神一振的建筑是一处环保教育的场所。

本土元素

材料是来自本土的：材料来自该中心所关注的环境，并和环境轻易地混合起来。比材料更重要的是其空间构成：建筑的大小相当于本地房屋的大小；有各个房间；中间有类似本地房屋的屋脊；四周的窗户从整体上把房子分割成小块。

现代元素

实际上这座本土化的建筑让人看起来有现代化特点的唯一因素是它的窗户，那些窗户没有任何支撑就相交成角。外行人会感觉窗户上方的石头就安置在窗户上。建筑的边界交角被隐藏起来了。

托肯湖游客中心

这座建筑就在瑞典森林的尽头和芦苇景区的起点迎接着游客。

本土元素

屋顶覆盖的茅草是就近取材的，这一材料的运用构成了本土化的元素。

现代元素

建筑的主要风格和形成这种风格的理念有着鲜明的现代化的意味。采用与地面和屋脊不平行的设计，使檐槽不规则地起落，形成抽象的建筑主体。墙体的交界处都是不规则的角度。由于这些出人意料形状和交界，内部空间的形态也没有规律，显得十分活跃。整座建筑像一个大飞船偶然降落此岛。在我们要如何憫视这一问题而去建造一座建筑以及如何建造一处容纳我们身体和心灵的空间的问题上，似乎难觅本土化的踪影。

behind our motivation to make buildings we can label as modern?

São Jorge Interpretation Center

This refreshingly modest building on São Jorge Island, Portugal, serves as a venue for environmental education.

What's vernacular;

The materials are clearly vernacular: they come from the environment that the centre is concerned with, and they blend in easily with that environment. But what is even more significant than the materials is the spatial composition: it has the scale of a vernacular house, with clearly defined rooms, a vernacular roof with the ridge in the middle, and window surrounds that form houses in miniature in the elevations.

What's modern;

Virtually the only element that turns this vernacular composition into something self-consciously modern is the windows that turn the corner without any visual support. To the untutored eye the stone above the windows rests on the window itself. The corner boundary of the building is in hiding.

Tåkern Visitor Center

This building greets its visitors at the point where a Swedish forest ends and a reed landscape begins.

What's vernacular;

The use of thatch for the roof covering is vernacular both as element and as material culled from the immediate environment.

What's modern;

The essence of the building, and the attitude that gives rise to it, are glaringly modern. Instead of running parallel to the ground and the ridge, the gutter rises and falls randomly, turning the building into an abstract mass. The walls are forbidden to meet each other at right angles. And interior spaces are active and busy due to their unexpected forms and boundaries. The whole composition might well be compared with spaceships that happen to have landed on the same field. They are hardly native to how we build without thinking about it, and how we build spaces to contain our body and our soul.

Dailai Conference Hall

A residential resort in Vinh Phuc Province, Vietnam, aims to bring

曼谷树屋的木质桥墩和桥，以及房间结构，使人们忆起一个充满乡情的世界

Bangkok Tree House's wooden piers, bridges and the structure of the rooms recall a vernacular world

黛莱会议中心

这是一处位于越南永福省的住宅区建筑，其设计目的是要把人们带回自然，这真的意味着我们在这座建筑里能获得当地生活的体验吗？

本土元素

进入会议大厅的路径把参观者带到并且穿过一个掩映在石墙中的优美的大门。学习建筑的学生立刻认出了这条路是来自于紫禁城的经典设计。进入大厅，参观者立即会被一片真正的当地竹林所倾倒，捆起来的竹子做成了桁架和房顶。竹子和石头都是本土化的元素，其设计理念来自Rudolfsky的"没有建筑师的建筑"。茅草屋顶为建筑加冕，是本土化材料中的一种。

现代元素

虽然桁架赋予大型室内空间一定的规模，但墙体却仍保持抽象的设计。石墙里整齐化一的洞使人们透过墙根本看不到整体的人形。除了大门以外，人们透过入口一侧的石墙也看不到房屋里的人形。桁架尽可能避免在屋内中央部分形成明显的屋脊，于是形成了不同的屋顶坡度。这样的设计明显来自现代理念，它告诉我们不应太保守。

曼谷树屋

该旅馆试图创造一种生活在树上的体验，即使房间并不是真的建在树枝上。

本土元素

木质的桥墩和桥、木质房间结构、房间和周围土地的密切关系：所有元素都使人们回忆起一个充满乡情的世界。

现代元素

设计本身意在掩饰组合和细节上的现代元素。可是有哪一个受过现代建筑学训练的建筑师能够完全摒弃现代化的设计方式呢？看看Aldo van Eyck的代表作品"阿姆斯特丹市立孤儿院"在构造方面的智能构成。又有哪一个设计本土化风格作品的建筑师会设计幕墙而放

people back to nature. Does that mean we can have a vernacular experience in it?

What's vernacular;

The route leading to the conference hall brings the visitor to and through a sensual gate punctured in an otherwise blind stone wall. Students of architecture would surely recognize in the route to and through the forbidden city. Once inside, the vistor finds himself captivated by a veritable forest of bamboo, bundled together to form the roof trusses and indeed the roof itself. Both the bamboo and the stone are vernacular materials collected in Rudolfsky's *Architecture without Architects*. A thatched roof crowns both the building and the list of vernacular materials.

What's modern;

Though the trusses give a certain scale to the large interior space, the walls remain abstract. The regimented holes between the stone blocks make it quite impossible to detect the scale of a human body in the abstract walls. Except for the gate, the stone wall on the entry side lacks all information that might let us see a body in it. The trusses do their best to avoid an obvious ridge in the middle of the space, resulting in different roof pitches. Such a design is clearly the result of a modern attitude that tells us we shouldn't be too traditional.

Bangkok Tree House

This hotel attempts to recreate the experience of dwelling in a tree house, even if the rooms don't literally rest on the branches of a tree.

What's vernacular;

The wooden piers and bridges, the wooden structure of the rooms, the relationship of the rooms with the land around them: they all recall a vernacular world.

What's modern;

The design itself belies the modern attitude both in composition and details. Which modern-trained architect would not recognize it in the structural and thus intellectual composition of Aldo van Eyck's Amsterdam Orphanage? Which vernacular builder would build curtain-wall elevations rather than windows cut in a wall? Even the stair with alternating treads tells us how strongly the architect was influenced by his training in the modern tradition.

BES Pavilion

In Vietnam a cluster of pavilions illustrates how we can build sus-

BES Pavilion, a sustainably built structure both in terms of materials and composition itself

弃在墙内嵌入窗户？即使是错步楼梯也告诉我们建筑师深受现代建筑理念的影响。

贝斯凉亭

在越南的许多凉亭设计中，我们可以看到利用建筑材料和建筑本身构成来推崇可持续理念。

本土元素

各种各样的单间凉亭随意分布在中心区域，仿佛它们是围绕神圣地方临时搭建的小屋，或者舞台上隔一段距离站好的演员。这些凉亭按照自然的秩序排列，毫不做作。土质或石质墙体、竹质横梁、竹片制成的屏风，所有这些本地材料构成了本地特色的设计。

现代元素

在这样小型的建筑中，很难找到现代建筑的痕迹。我们可以认为墙是现代化的。墙体很厚重，像永不走动的飞机。在这方面有一些关于构成的认知主义的建造方法被掩盖了。

兰溪庭

兰溪庭是中国成都的一所私人俱乐部，还包括一个餐馆和一个内部庭院，是用现代视角来解读中国传统园林和建筑的一次尝试。

本土元素

门廊所用的栋和梁、有着优美弧度的屋顶，无不表现了紫禁城的传统设计对建筑师的启示。百叶窗也并非是永恒的。其空间设计曲径通幽，又别有洞天，这同样体现了来自中国传统园林的设计灵感。由于人是整座建筑整体的创建者，所以这些设计无疑是本土化的。

现代元素

虽然设计师受到中国传统园林的影响，其设计理念却是现代的、概念性的。设计师想尝试创造一些新的、与众不同的风格，而不是只是客观建造有意义的、好的事物。为了再现当地的水景，设计师设计了一种波状的斜屋顶。这种寓言式的符号在本土化建筑里非常少见。本土化建筑的设计者为人类的身体和心灵建造了家园。墙体设计有点写

tainably, not only with materials but also with the composition itself.

What's vernacular;

The various one-room pavilions stand informally around a center space as though they were huts grouped around holy ground, or else actors standing at exactly the right distance from each other on a stage. They embody a natural order that does not seem all contrived. The walls made of either loam or stone, the beams made of bamboo, and the screens made of smaller pieces of bamboo: they are all native materials that work together to form a composition native to how we can build.

What's modern;

You have to search quite hard to find a residue of the modern attitude in this gentle building. You might call the walls modern. Though they're undeniably massive, they stand as planes that never turn the corner. In this respect they belie a certain cognitive approach to the composition.

Lanxi Curtilage

A private club in Chengdu, China, including a restaurant and an inner courtyard, is an attempt at a modern interpretation of a traditional Chinese garden and architecture.

What's vernacular;

The vernacular discoveries in the Forbidden City have inspired the architects here as well, both in the columns and beams that form one of the porches and in the gracefully curved roof above it. The window shutters are unattestably timeless as well. And the spatial composition juxtaposes small spaces with larger spaces and then with still larger spaces. The inspiration is said to be a traditional Chinese garden. The result is in any case vernacular since the human body is the generator of the spatial building blocks.

What's modern;

The attitude the architects took when they were inspired by the traditional Chinese garden is, however, modern. It's conceptual. It's the attempt to create something new and different rather than the attempt to build something objectively good and meaningful. This same modern attitude led the architects to create an undulating series of pitched roofs in order to recall the region's rivers. Nowhere in vernacular architecture do builders try to create symbols which become no more than allegories. Vernacular builders create homes for the human body and the human soul.

兰溪庭的波状墙体,灵感来自于对水(一种灵活的自然理念)的解读

ripple wall of Lanxi Curtilage which was derived from an interpretation of water, a flexible natural conception

肉桂培训中心非常具有本土特色,有永恒的感觉

the spatial composition of Cassia Coop Training Center is vernacular in the timeless sense

实,有点抽象,有点无形,它们给来访者一个信息,而不是一个拥抱。然后郑重地告诉来访者,一旦住在这里,就要与周围联系在一起,而庭院四周的幕墙里之间的联系或是利华大厦,或是西格拉姆大楼。

肉桂培训中心

在苏门答腊岛推崇环保的肉桂学校,工人们学习肉桂交易,同时也了解了他们拥有的权利。建筑物中间的柱子都是通常被扔掉的肉桂树制成的。

本土元素

空间设计非常具有本土特色,有永恒的感觉:空旷的广场周围是一些小的建筑,它们是整座建筑的中心。这座建筑采用一个人体比例的规模,站在那里,适得其所,又像若干位演员在舞台上讲着一个故事。

除了混凝土,建筑材料都来自当地:本地制作的砖和肉桂树的树干制成的柱子。建造方法也是用当地建造房子的方法:由那些当地的、没有受过培训的工人用双手建造。建筑本身就告诉你这是一个手工建造的房子。

现代元素

窗户是随意嵌入墙内的,毫无章法,让我们想起那些获奖的当代建筑中精心设计的窗户。柱子也随着自己的形状,与倾向于定义的空间主题无关。我们在这里找不到人为的感觉。如果我们是专业建筑师,你可以找到勒·柯布西耶的底层架空的设计。底层架空的设计对空间没有限定。隔离房间的墙体和支持房顶的结构是分开的,这种设计有柯布西耶式的,有现代的,还有意识形态的痕迹。大窗户的窗格没有直角的,这一设计灵感也来自当代设计风格。这些设计与我们对于墙、空间和对我们自己的认识毫无关联。

现代和本土的合成体,可能吗?

建筑和空间在和我们对话。它们告诉我们,我们建造了什么。它们体现了我们在设计和建造这些建筑作品时的态度。也告诉我们如何

The walls of the building are similarly cognitive, similarly abstract, similarly scaleless: rather than embrace the visitor, they give him a message. The next step would be to hand out folders telling visitors what they were meant to associate once they inhabited the building. And the only association possible in the curtain walls surrounding the courtyard is either Lever House or the Seagram Building.

Cassia Coop Training Center

In this sustainable cinnamon school in Sumatra, workers learn both their trade and their rights. And the columns in the building are the parts of the cinnamon trees that are normally discarded.

What's vernacular;

The spatial composition is vernacular in the best, timeless sense: small buildings cluster together to enclose an open square. They make a centre of it. The scale of the buildings is the scale of a human body. The buildings stand in just the right places as though they were human actors telling a story on stage.

Except for the concrete, the building materials are vernacular as well: locally made bricks and native cinnamon trees as columns. The building method is vernacular too: local, unskilled people build with their own hands. The building itself tells you it was built with human hands.

What's modern;

The windows are haphazardly punched in the walls: they form no meaningful composition; they remind us of the arbitrarily placed windows in many prize-winning contemporary buildings. The columns follow their own path, quite divorced from the spaces they help to define. We can't recognize a human body in them. If we are trained architects we might recognize Le Corbusier's piloti in them: piloti plays no role in defining space. The walls that define the rooms are separate from the structure that supports the roof: equally Corbusian, equally modern, equally ideological. And the panes in the larger windows are not allowed to be orthogonal: clearly another borrowing from current modern architecture. They have nothing to do with how we perceive walls, how we experience space, how we regard ourselves.

Modern vernacular: Is a synthesis possible?

Buildings and spaces speak to us. They tell us what we've built. They embody the attitude we took when we designed and built them. And so they tell us how we use that attitude and how that

利用这种态度,以及态度是如何利用我们的。
　　希望建造本土现代化建筑的意图本身就是一种态度。这种态度是现代建筑的核心教义之一：建造现代的建筑；建造只属于我们这个时代的建筑。在建筑的世界里只有一个上帝,他的名字叫做"时代精神"。
　　建造本土化的建筑不是要跟上潮流。它也不是教条。我们要建造一座能让我们联想起山脉、波涛或是森林的建筑并不难。本土化的建筑风格来源于我们本身对空间的感受。我们渴望有一处空间能容纳我们,我们想看到自己在周围的空间里的反映。我们需要一处生气勃勃的空间,而不是举目只有没有生命的物件。
　　我们设计和建造所使用的工具都是我们双手的延伸。难道那意味着电子技术是当代的方言吗?
　　答案不在工具本身,而在我们使用它们时的态度。我们究竟要设计和建造什么呢? 我们意识到我们所希望的了吗? 我们的希望比我们现在意识到的要深奥多吗?
　　从本土化的建筑来看,确实是这样。一个本土化的世界就是当地人的世界,一个我们知道自己是谁且在做什么的世界。本土化的世界充满了"意义",但不是我们所认为的"意义",而是我们碰到和认可的意义。一旦邂逅了这个"意义",你不必去特意编写它,把它制成教条。一个梦或是一个景象会直指你心,而一个教条则只能让你记住而已。我们建造的世界是个制造麻烦的世界。
　　我们这个特定的时代不神圣,而生活是神圣的。我们能够建造一个我们觉得神圣且统一的世界。当我们这么做了,我们就不必问我们的建筑是现代化的还是本土化的。我们就不再执着于我们作为建筑师应该怎么样,我们的同事怎么看待我们。我们只是建设者,永恒的建设者,我们有天生的使命,那就是建造一个符合我们天性的世界。

attitude uses us.
The attempt to create a building we might call modern vernacular is itself the result of an attitude. That attitude is one of the cardinal doctrines of modern architecture: building something we can call modern; building something that belongs exclusively to our own age. It is as though there was but one god in the world of architecture, and the Spirit of the Age is his name.
The attitude that produces vernacular architecture is not concerned with keeping up with the times. It is not doctrinal. It doesn't challenge us to make a building that reminds us of a mountain range or a wave or a forest. The vernacular attitude comes from the knowledge our body has of spaces. We yearn for spaces that contain us. We want to see ourselves reflected in the boundaries of those spaces. We long for spaces we can experience as alive, not spaces we encounter as things.
The tools we use when we design and build are extensions of our hands. Does that mean that digital technologies are the vernacular of our current age?
The answer lies not in the tools but in the attitude we take when we use them. What, really, do we want to design and build? Are we conscious of our wishes? Are our wishes deeper than what we think about them?
The record of vernacular architecture proves they are. A vernacular world is a native world, a world native to who we are and what we are. A vernacular world is a world filled with meaning, not meaning we think about, but meaning we encounter and recognize. Once you've met meaning, you no longer have to codify it, to make a doctrine of it. A dream or a vision touches all of you. A doctrine touches only your head. And in the world we build, it produces headaches.
Our particular age is not holy. Life is holy. We are capable of building a world we can experience as holy, as whole. When we do, we no longer ask whether we can call our architecture modern vernacular or not. We're no longer preoccupied with who we are as architects, with what our colleagues think of us. We're simply builders, timeless builders, with an inborn task: to build a world native to our own nature. Jaap Dawson

本土现代化 | Vernacular and Modern

圣乔治环境展览中心
Ana Laura Vasconcelos

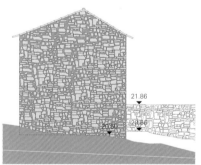

东立面_公寓 east elevation_apartment

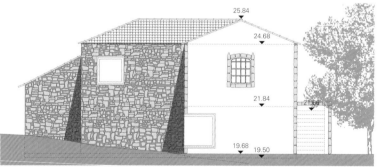

东立面_环境展览中心 east elevation_environmental interpretation center

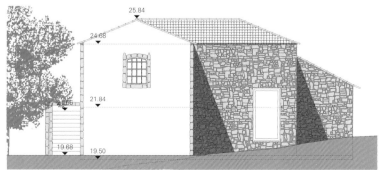

西立面_环境展览中心 west elevation_environmental interpretation center

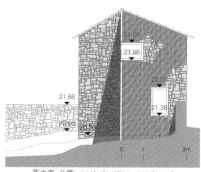

西立面_公寓 west elevation_apartment

环保的环境展览中心将整个区域连接起来,并且处在其控制的范围之内。这个中心也包括一处即将在Faja da Caldeira de Santo Cristo建立起来的营地。因此,它成为第一处信息联系中心,能够公开现场信息,并且在这处价值较高的场地发挥着实质的功能,即对环保价值进行诠释,它们包括:

——简化和丰富Faja da Caldeira de Santo Cristo区域的访问路线。

——唤醒与景观和乡村价值的展览相关的环保教育。

——创造具有诠释性的自然路线和建筑遗产。

——开拓海洋和沿海栖息地,培养多样化的动植物的知识。

整个项目包含两座建筑:环境展览中心以及一直支持研究(根据原始规划进行了重建)的公寓。周围的石墙以及现存的饮水资源得以保留,并且进行重新定义。

第一座建筑,即环境展览与监控中心,包含两个楼层,成为现代与传统之间的互动。主体量为"T"形(在Faja地区传统的房屋建筑中较为普遍),再现了原始立面的特点和设计,且保留了推拉窗和门,包括材料,无论它们是不是由当地的木框还是玄武岩石制成的门框和横梁,都加以保留。垂直的体量将立面和石质烟囱保持在视线范围之内,同时引进一些现代元素,使其与营地之间的连接即将产生。

第二座建筑被规划为一座临时公寓,由于其规划的功能而变得现代化,且具有适应性。它将两个楼层结合起来,来作为一处开放的空间。地面层容纳了厨房和起居室,而二层则包括浴室和卧室。

São Jorge Interpretation Center

The environmental interpretation center articulates and controls the entire area. It will also include a campsite which will be established in Faja da Caldeira de Santo Cristo. It is thus, the first contact information intended to disclose "in situ" and shall have an essential function, the explanation of the environmental values at this location is of high value, namely:
– streamline and enrich the visitation of Faja de Santo Cristo;
– awake the environmental education related to the interpretation of the values of landscape and countryside;

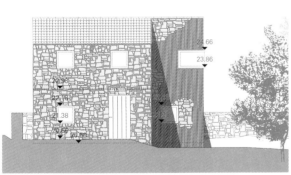

北立面_公寓 north elevation_apartment

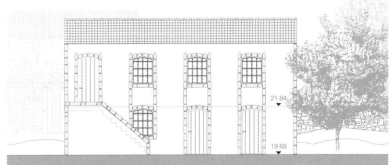

北立面_环境展览中心 north elevation_environmental interpretation center

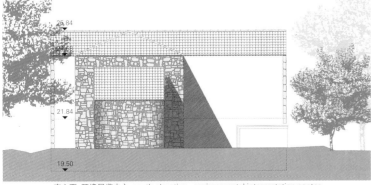

南立面_环境展览中心 south elevation_environmental interpretation center

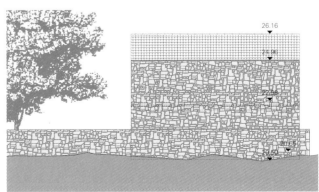

南立面_公寓 south elevation_apartment

项目名称：Environmental Interpretation Centre in São Jorge Island
地点：Fajã da Caldeira de Santo Cristo, São Jorge Island, Açores
建筑师：Ana Laura Vasconcelos
工程师：Batiaçores, Lda
结构工程师：Helena Batista
机械工程师：João Paulo Veloso
照明工程师：João Mota Vieira
甲方：Government of the Açores
用地面积：1,275m² 总建筑面积：256m² 有效楼层面积：280m²
造价：EUR 347,862.79
设计时间：2007—2008 施工时间：2009—2011
摄影师：©FG+SG Architectural Photography

现存建筑，1980年 existing building, 1980

现存建筑，2007年 existing building, 2007

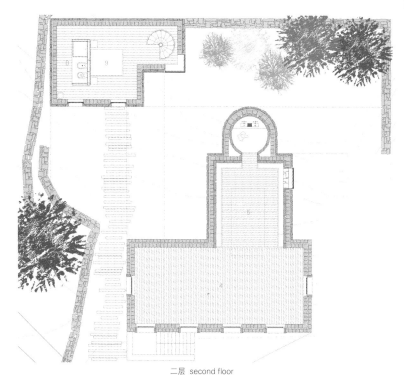

二层 second floor

一层 first floor

1 售票处
2 礼品店
3 入口
4 展览区
5 音像图书馆
6 起居室
7 厨房
8 卫生间
9 卧室

1. ticket office
2. gift shop
3. entrance
4. exhibition
5. video library
6. living room
7. kitchen
8. toilet
9. bedroom

环境展览中心
environmental interpretation center

– create interpretive natural routes and architectural heritage;
– develop knowledge of marine and coatal habitats, as well as its diverse fauna and flora.

The entire project consists of 2 buildings: the environmental interpretation center and an apartment to support research, reconstructed according to the original plans. The surrounding stone walls were kept and redefined, as well as the existing drinking water sources.

The first building, the Center for Interpretation and Environmental Monitoring, consists of two floors, and is an existing interaction between "modern" and traditional. With the shape of a "T" (architecture prevalent in traditional houses of Faja), the main body relives the features and design of the original facade, keeping the sash windows and doors of stature, including the materials, whether it be the wood frames or the doorposts and beams in basalt stone from the region. The perpendicular body keeps the facades and stone chimney in sight, introducing, however, some contemporary elements, so the connection with the campsite is forthcoming.

The second building, intended for a temporary apartment, is modernized and adapted according to the new functions which it is intended for. It incorporates 2 floors, functioning almost as an open-space: Floor 1 consists of the kitchen and living room and the 2rd floor consists of the bathroom and bedroom.

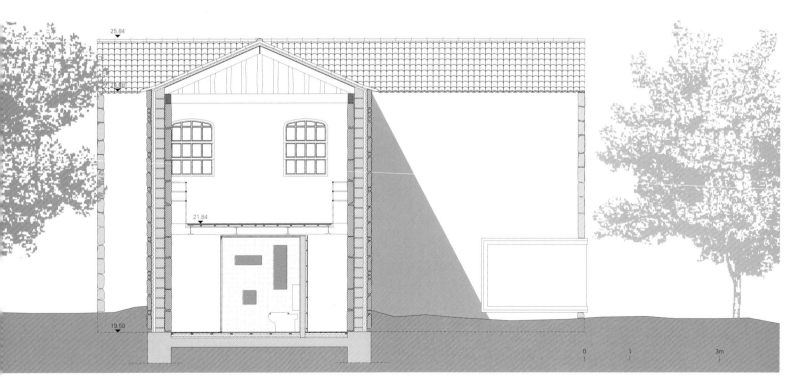

A-A' 剖面图_环境展览中心 section A-A'_environmental interpretation center

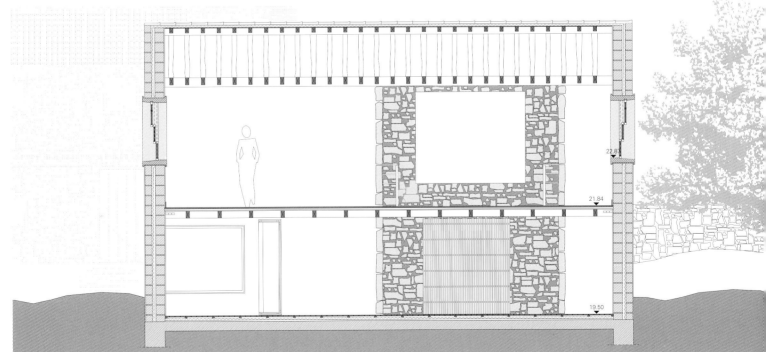

B-B'剖面图_环境展览中心 section B-B'_environmental interpretation center

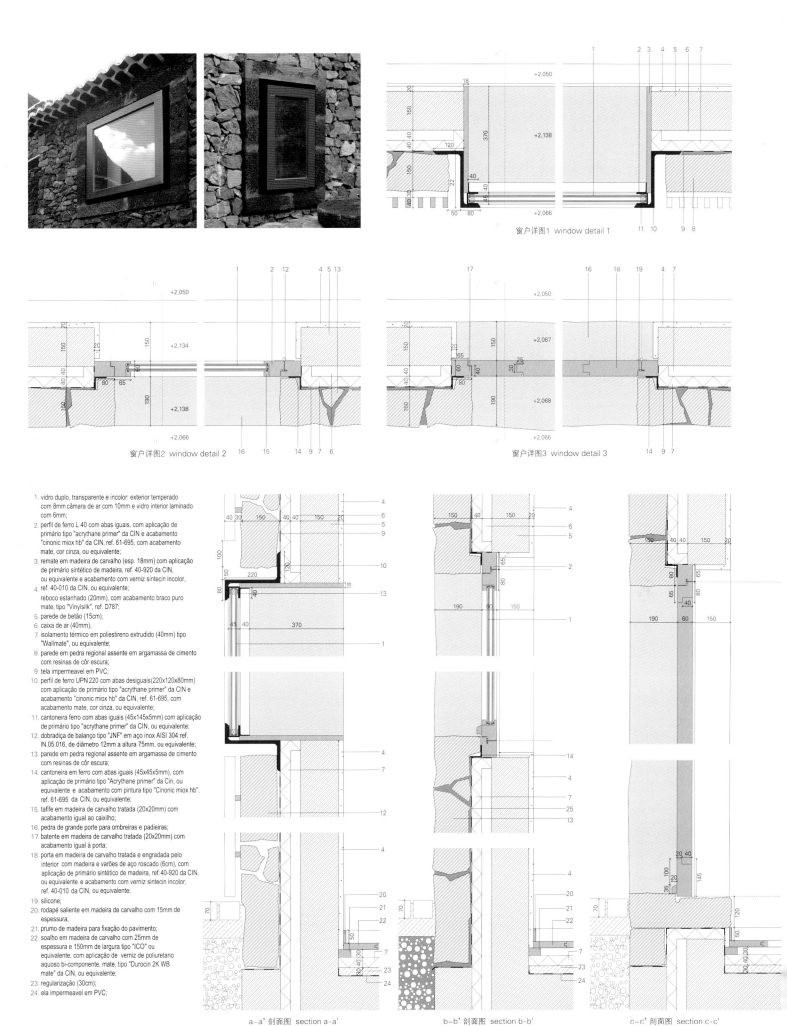

托肯湖游客中心
Wingårdh Arkitektkontor AB

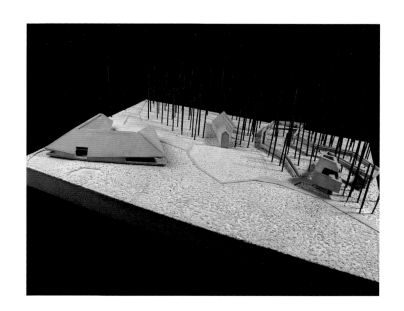

这些建筑的设计目标是增强游客的大自然体验，建筑师的设计和建造成为这个地方非常迫切的需求。托肯湖鸟类保护区有大片的湿地芦苇区、集体水下生活区、广阔的海滩草地，以及湖畔沿岸的开放式森林，不仅仅对于鸟类，还对于那些来此享受自然赐予的财富的人们来说，都是一处天堂。

托肯湖游客中心第一眼看上去像是一个从芦苇中削减出来的立体体块。但是实际上，茅草覆盖的屋顶呈折叠状，以形成一处面向鸟类和天空开放的室外空间。折叠产生了有机形式，使其与周围环境非常融洽。这座中心及其瞭望塔被规划得比森林边缘处的隐藏地带更大一些。

位于森林边缘的场地允许建筑和瞭望塔安置在树林和芦苇中。曲折的小径穿过森林，带领游客到达瞭望塔，如同建筑的庭院遮蔽处一样，使人们免受大风的侵袭。小径不仅仅是一条让人们倍感舒适的道路，从上面看，它还是一处供人们一览湖面风景的场地。

主建筑的外观不仅仅满足了室内最小洞口的一系列要求。展览空间倾向于建成封闭的空间，但是也需要来自上方的日光。此外，除了位于屋脊上的天窗，建筑还要从三个不同的点来获取阳光和视野。第一处为入口，它连接着室外的庭院，使通往展览区的道路十分便捷，也十分舒适。第二处洞口沿着入口的轴线设置，引领游客进入展览空间。第三处洞口则位于空间偏远的角落里，细长且低矮的窗户为人们提供了全景。

屋脊是茅草屋顶最脆弱的地方。在这里，茅草被玻璃所取代，这不仅仅是一种安全的技术方法，来对茅草工程进行收尾，还使室内完全沐浴在天空的阳光下。

抬头向上看，人们能够看见这只展出的木鸟（建筑），天空在后面成为其自然背景。

通过这种方式，托肯湖的芦苇不仅成为建筑的覆层材料，而且还对其形式产生了重要的影响。茅草是一种实际的建筑材料，如果损坏的话，可以较为容易地进行翻新。屋顶和墙体在20cm厚的保温层上覆盖了超过25cm厚的茅草，形成了一处舒适的庇护所。斜屋顶赋予茅草屋顶超过50年的寿命。建筑所应用的本地材料不仅仅包括托肯湖的一小部分芦苇。设计通过升级具有明显现代形式的本地传统茅草屋顶，来与历史进行对话。而另一次对话则是建筑的合理形式与其经典剖面之间的对比。这是一座为时间而设计房间的建筑。

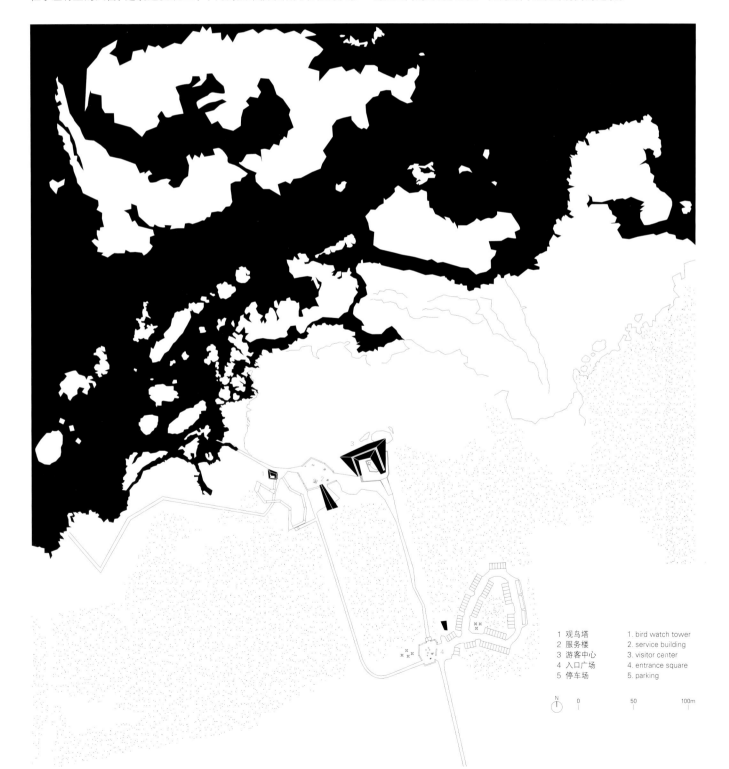

1 观鸟塔　　1. bird watch tower
2 服务楼　　2. service building
3 游客中心　3. visitor center
4 入口广场　4. entrance square
5 停车场　　5. parking

Tåkern Visitor Center

The goal for these buildings is to heighten the visitors' experience of nature, and that place greatly demands on both their design and their execution. The Lake Tåkern Bird Sanctuary, with its enormous fields of wetland reeds, teaming underwater life, broad beach meadows, and open woods along the water's edge, is a kind of paradise – and not just for the birds, but also for those who come to enjoy the wealth that nature has to offer.

The Tåkern Visitor Center looks at first like a solid block cut from the reeds. But the thatched building is actually folded to form an outdoor room open to the birds and the sky above. The folds generate organic forms that are at home in their natural surroundings. And the center and its observation tower are intended to be little more than hiding places at the edge of the forest.

The site at the forest's edge allows the building and its tower to be

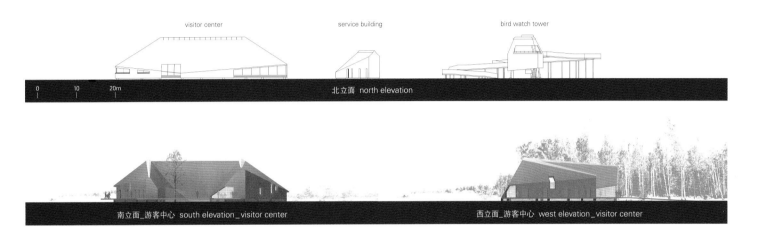

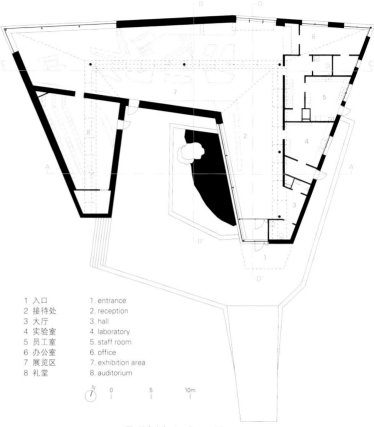

1 入口	1. entrance
2 接待处	2. reception
3 大厅	3. hall
4 实验室	4. laboratory
5 员工室	5. staff room
6 办公室	6. office
7 展览区	7. exhibition area
8 礼堂	8. auditorium

一层_游客中心 first floor_visitor center

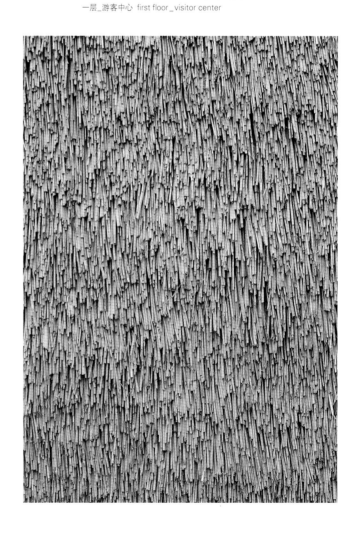

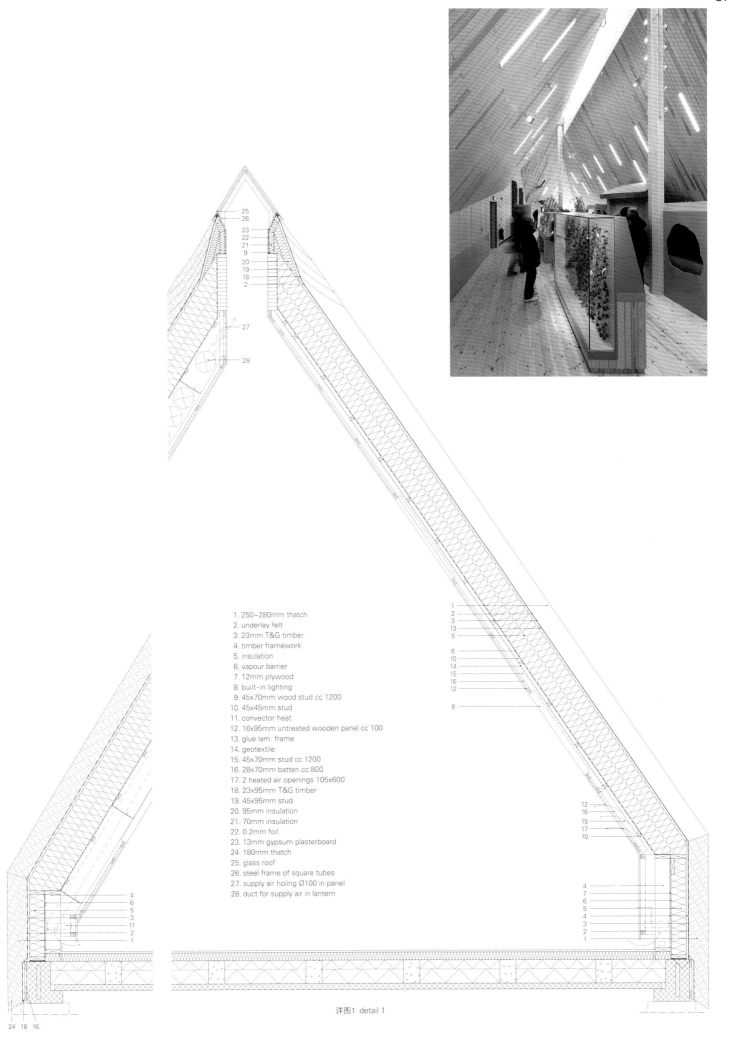

1. 250~280mm thatch
2. underlay felt
3. 23mm T&G timber
4. timber framework
5. insulation
6. vapour barrier
7. 12mm plywood
8. built-in lighting
9. 45x70mm wood stud cc 1200
10. 45x45mm stud
11. convector heat
12. 16x95mm untreated wooden panel cc 100
13. glue lam. frame
14. geotextile
15. 45x70mm stud cc 1200
16. 28x70mm batten cc 800
17. 2 heated air openings 105x600
18. 23x95mm T&G timber
19. 45x95mm stud
20. 95mm insulation
21. 70mm insulation
22. 0.2mm foil
23. 13mm gypsum plasterboard
24. 180mm thatch
25. glass roof
26. steel frame of square tubes
27. supply air holing Ø100 in panel
28. duct for supply air in lantern

详图1 detail 1

in the woods and among the reeds as well. The meandering path through the woods to the tower takes care of its visitors, just as the building's courtyard shelters them from the wind. The path is more than just a suitable way to get a view of the lake from above. The look of the main building is more a consequence of the interior's minimal need for openings. Exhibits prefer enclosed spaces, but also daylight from above. In addition to skylights at the ridge, the building needed light and views from three different points. The first is the entrance, where the contact with the courtyard outside makes the way into the exhibition short and pleasant. The second opening is a view out along the entrance axis that leads visitors into the exhibition space. The third is in the far corner of the space, where a long and low window offers a panoramic view. The ridge is the most vulnerable part of a thatched roof. Here the thatch has been replaced by glass, which is not only a technically safe way to finish off the thatch but bathes the interior in light from the sky.

Looking up, people see the wooden birds of the exhibition with the sky behind them as in nature.

In this way, Lake Tåkern's reeds give the building not just its cladding but much of its form. Thatch is a practical material that can easily be renewed if it should be damaged. The roof and walls are clad with more than 25cm of thatch over 20cm of insulation, and together they make for a cozy refuge. The steep pitch of the roofs gives the thatching an estimated lifespan of more than 50 years.

A small portion of the lake's reeds are not the only local material used in the building. The design establishes a dialogue with history by updating the local tradition of thatched roofs with a decidedly modern form. Another dialogue is the contrast between the building's rational form and its classical section. It's an architecture that makes room for time.

项目名称：Naturum Tåkern
地点：Glänås, Sweden
建筑师：Wingårdh Arkitektkontor AB
首席建筑师：Gert Wingårdh, Jonas Edblad
项目团队：Ingrid Gunnarsson, Aron Davidsson, Jannika G Wirstad, Peter Öhman, Danuta Nielsen, Björn Nilsson
项目管理：TP Group/Sundbaums Byggprojektering AB
结构工程师：Cowi
HAVC工程师：Bengt Dahlgren AB
电气工程师：WSP
消防工程师：Bengt Dahlgren AB
音效工程师：Akustikforum AB
建造商：Skanska Sverige AB, Håkan Ströms Byggnads AB
开发商：County of Östergötland
业主：Naturvårdsverket
有效楼层面积：750m^2
施工时间：2008—2012
摄影师：
©Tord-Rickard Söderström (courtesy of the architect)
-p.30^{middle}, p.30~31^{bottom}, p.34~35, p.37, p.39
©Åke E:son Lindman (courtesy of the architect)
-p.30^{top}, p.32~33, p.35, p.36, p.38 (except as noted)

三层 third floor

一层 first floor
观鸟塔 bird watch tower

二层 second floor

E-E' 剖面图 section E-E'

F-F' 剖面图 section F-F'

一处名为黛莱火烈鸟度假区的住宅区被规划起来，且部分是为繁忙的城市居民而建，使他们能够享受到自然环绕的周末时光。场地位于黛莱湖与周围群山之间的繁茂的森林中，距离首都河内约50km远。度假区的访客也可享受内嵌各种自然物、植物和花朵且风景优美的景观，使人们逃离日常生活的狭小区域。

黛莱会议中心的场地位于主路的一侧，曾作为整个度假区的入口，当游客到达此地时，建筑呈现出欢迎的姿态。为了提高游客来此度假区享受愉快时光的期望，建筑师沿着道路设计了一面令人印象深刻的曲形石墙，来迎接游客。这面墙体80m长、8m高、1m厚，为游客提供了一系列的景色，展示了周围各处的自然风景。此外，曲形墙体也作为一个设施，提高了游客的士气，同时试图将他们吸引到大厅内举办的活动中来。一个垂直入口位于人工小山之间，引导游客穿过石墙，到达充满活力的竹结构(超大规模)覆盖的门厅中。

会议中心大跨度的结构材料包含直挺的竹子，竹子本身有许多的优点，如赏心悦目的颜色、质地以及其再生的可能性。大量的竹子形成一个结构框架，会比单个的竹子更加具有稳定性。这个结构的最大跨度为13.6m，每个框架的节点位置都加以调整，来创造一个宽大的屋顶曲面。尽管会议中心的功能要求将空间分割成特定的房间，如主厅、附属厅、门厅以及后备室，但是其充满活力的竹结构仍然能使游客感受到空间是更加开阔、更加开放的，并且还通过隔墙上方的气窗来形成连续性。

竹子和石材是附近区域较为富足的自然资源。通过大量采用这些当地的材料，这个大厅具有了原始的特色，且形成了特殊的氛围。因此，建筑与自然友好共存。这座建筑不仅仅旨在提供一处举办活动的绝佳场所，同时还要深化广袤自然精髓所带来的体验。

Dailai Conference Hall

A residential resort, named Flamingo Dailai Resort, was planned and partly constructed for busy city citizens to enjoy their weekends surrounded by nature. It is located in the middle of flourishing forests between Dailai Lake and surrounding mountains, about 50km away from Hanoi. The guests of this resort can enjoy the beautiful landscape inlayed with numerous natural objects, plants and flowers and escape from their daily life in cramped quarters.

The lot of Dailai Conference Hall is located beside the main access road, which is used as an entrance for the whole resort; the building welcomes all visitors when they come. To enhance their expectation for a delightful stay in the resort, an impressive curved

黛莱会议中心
Vo Trong Nghia+Takashi Niwa

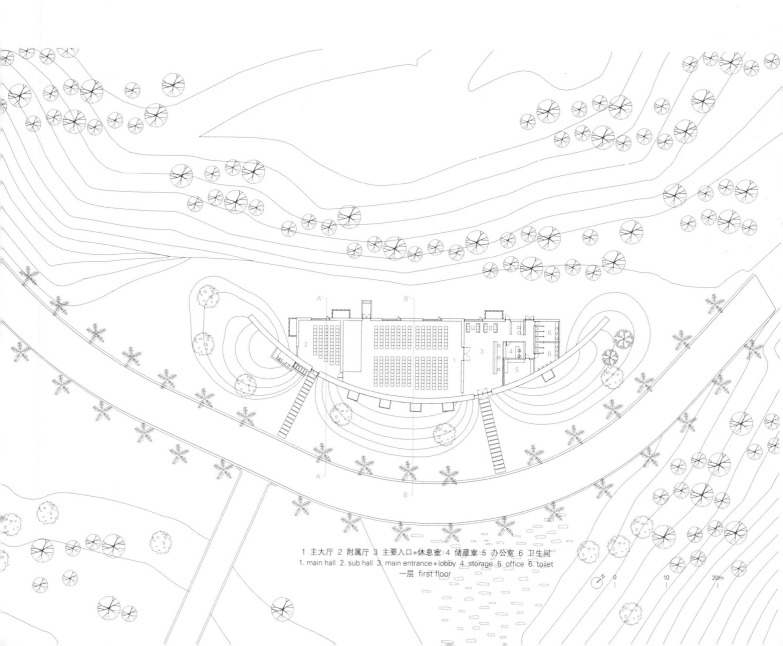

1 主大厅 2 附属厅 3 主要入口+休息室 4 储藏室 5 办公室 6 卫生间
1. main hall 2. sub hall 3. main entrance+lobby 4. storage 5. office 6. toilet
一层 first floor

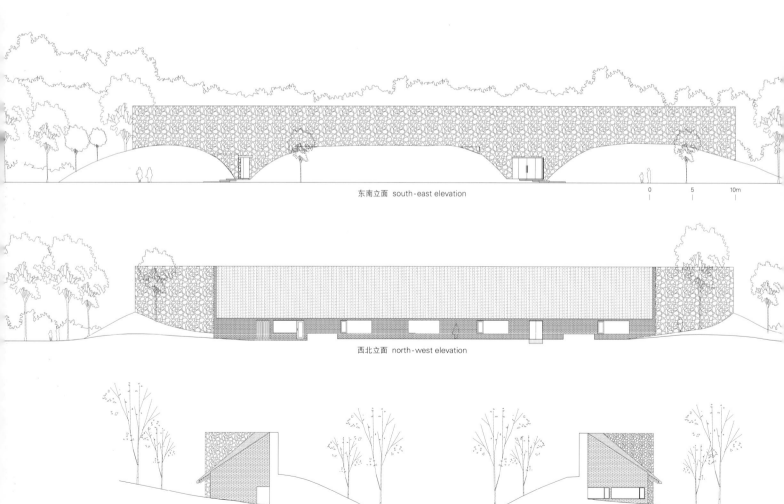

东南立面 south-east elevation

西北立面 north-west elevation

西南立面 south-west elevation

东北立面 north-east elevation

stone wall along the road was designed as its eceptionist. The wall, which is 80-meters long, 8-meters high and 1-meter thick, offers a sequential view to visitors, revealing and screening the surrounding nature from place to place. Furthermore, the curved wall works as a device, which raises the morale of visitors and attempts to lure them to events being performed in the hall. An orthogonal access between artificial hills conducts visitors through the stone wall, then, visitors reach a foyer covered by a dynamic bamboo structure with an extraordinary scale.

The wide-span structure of the conference hall consists of the composition of straight bamboos. Bamboo itself has many advantages, such as beautiful color, texture and reproduction potential. Many bamboos are assembled into a structural frame, which has higher reliability and redundancy than bamboo used individually.

Its maximum span is 13.6meters and the positions of the joints at each frame are adjusted to make a generous curve of the roof. Though the functional requirement as a conference center divides the space into specific rooms such as the main hall, sub hall, foyer and supporting rooms, the dynamic bamboo structure enables visitors to feel that the spaces are wider and more open, showing its continuity through a transom window above the partitions.

Bamboo and stone are abundant natural resources near the area. The hall achieves its originality and special atmosphere by using these local materials in plenty. Consequently, the building becomes a friendly accompaniment to nature. The aim of this building is not only to supply a nice space for events but also to deepen the experience of the generous spirit of nature.

项目名称：Dailai Conference Hall
地点：Flamingo Dailai Resort, Vinh Phuc Province, Vietnam
建筑师：Vo Trong Nghia, Takashi Niwa
承包商：Hong Hac Dailai JSC, Wind and Water House JSC
甲方：Hong Hac Dailai JSC
功能：conference hall
用地面积：2,700m² 有效楼层面积：730m²
设计时间：2010.6 施工时间：2011.1 竣工时间：2012.8
摄影师：©Hiroyuki Oki (courtesy of the architect)

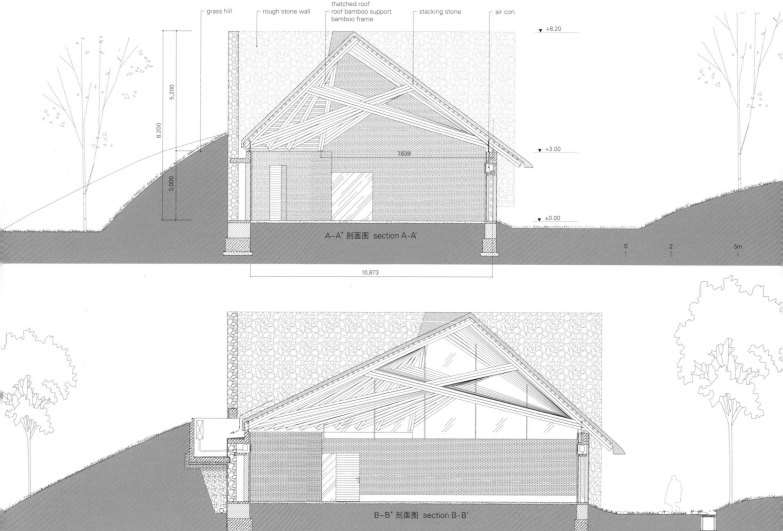

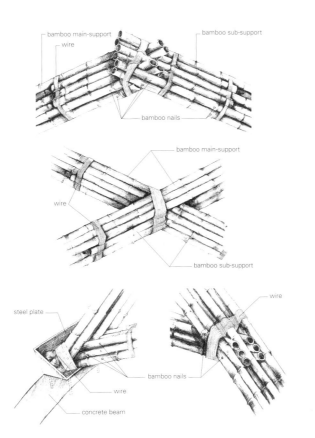

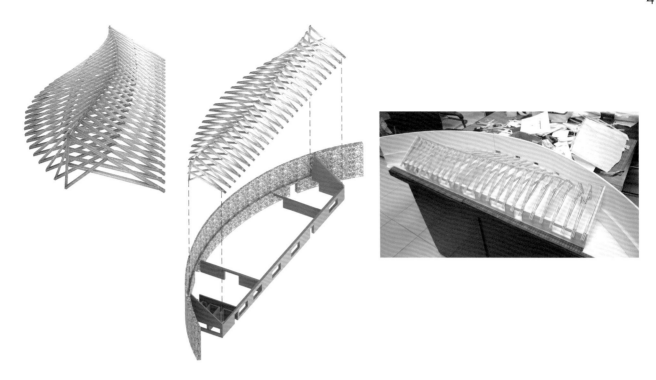

本土现代化 Vernacular and Modern

曼谷树屋

Nuntapong Yindeekhun + Bunphot Wasukree

时间机器穿梭到未来
主要的设计方法起源于泛舟的河流/主要的楼层由长钢杆支撑,拔地而起/使地面层远离地面/
让水来回地奔腾/让爬行动物和当地树种立足于地面/
人们攀爬至屋顶露台,来拥抱天空/
蜿蜒的客房/创造了各具特色的景观视野/
蜿蜒的廊道/一望无尽头/
竹制的新装/再生的木材,与水结伴而行/
我们以玻璃镜来武装,削减了体量/披上绿色的树衣/赋予你双生景象

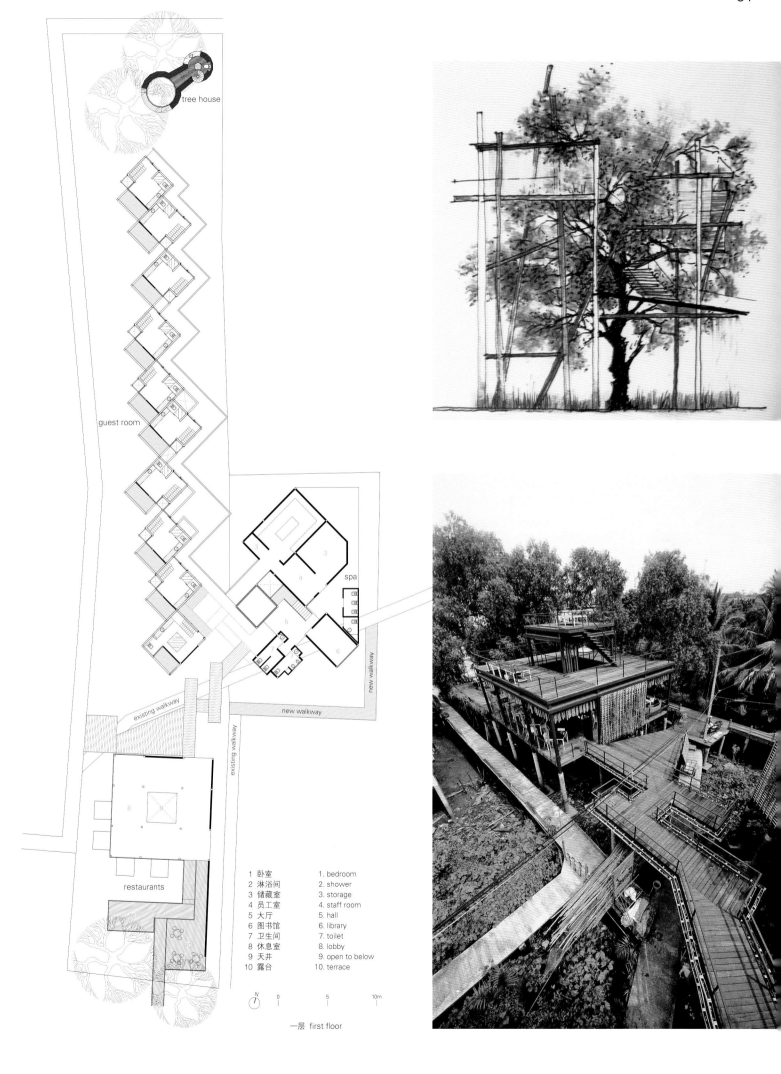

Bangkok Tree House
The time machine to the future

Main approach from river by boat / main floor raised from earth by long steel pole / leave ground to the earth /
Let water come up and down / let reptiles and local trees stay on their ground /
Open to sky by roof terrace that you have to climb up /
With zig-zag module guest bedroom / creat each room with their own view /
With zig-zag corridor / you will see no end /
Skin with local bamboo / and recycle wood that come with the water /
We erase some mass with mirror / and give more tree on that / you will see twin trees /

Nuntapong Yindeekhun + Bunphot Wasukree

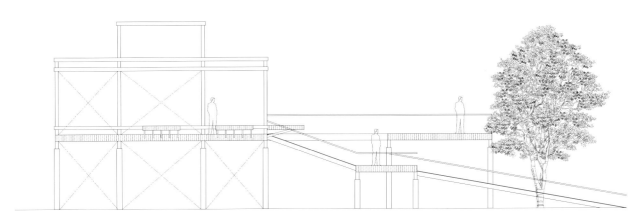

A-A' 剖面图 section A-A'

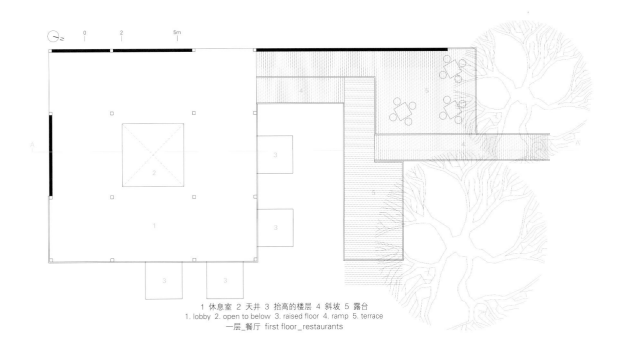

1 休息室 2 天井 3 抬高的楼层 4 斜坡 5 露台
1. lobby 2. open to below 3. raised floor 4. ramp 5. terrace
一层_餐厅 first floor_ restaurants

1 露台 2 淋浴区 3 卫生间 4 厨房 5 大厅 6 等候区 7 楼梯
1. terrace 2. shower 3. toilet 4. kitchen 5. hall 6. waiting area 7. stair
一层_水疗区 first floor_spa

西北立面_水疗区 north-west elevation_spa

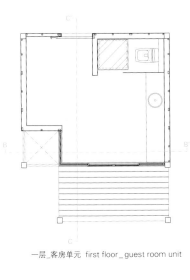
一层_客房单元 first floor_guest room unit

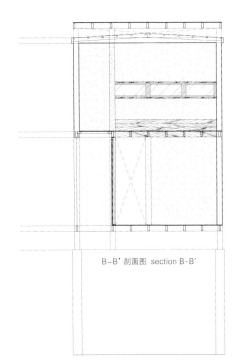
B-B' 剖面图 section B-B'

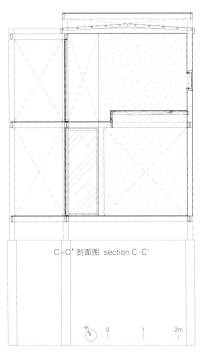
C-C' 剖面图 section C-C'

0　1　2m

项目名称：Bangkok Tree House
地点：Bangkok, Thailand
建筑师：Nuntapong Yindeekhun, Bunphot Wasukree
结构工程师：Suwatcha Yindeekhun
用地面积：1,350m² 有效楼层面积：670m²
竣工时间：2011
摄影师：
Courtesy of the architect - p.52~53, p.56, p.57
©Bangkok Tree House - p.50, p.51, p.53, p.54, p.55, p.58, p.59

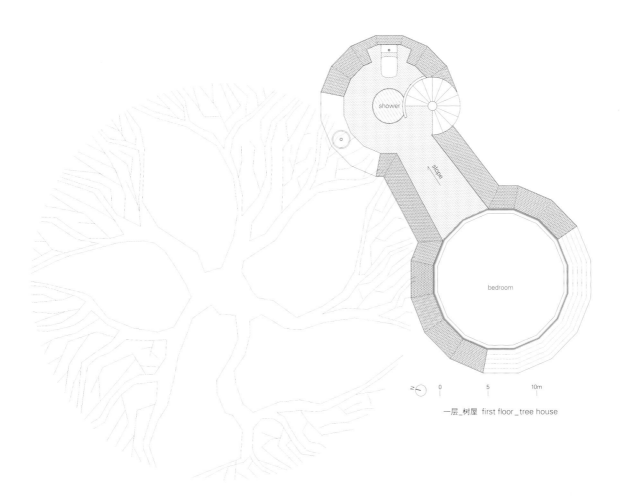

一层_树屋 first floor _ tree house

贝斯凉亭
H&P Architects

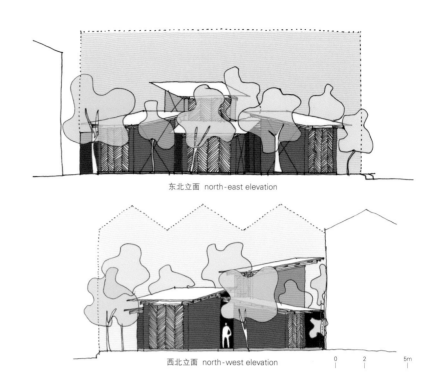

东北立面 north-east elevation

西北立面 north-west elevation

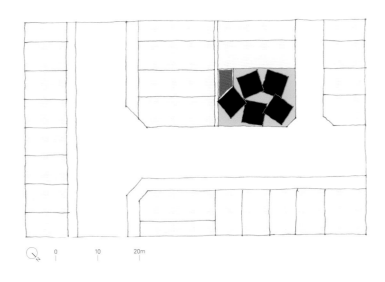

贝斯凉亭是一处开放的社区服务空间，集中展示艺术和文化的各个方面。贝斯凉亭（竹子+土+石材）位于越南河静市的中心，基于以用户为中心的设计理念，采用当地的建筑方法，由当地材料建造起来。

这一建筑集群包括各式各样的隔间，它们围绕着中心庭院自由地布局，以产生不同的景观视野，同时光与影的互动也在此形成。这些都有助于消除室内外之间的界限。

这座建筑的使用者们都极有可能对其功能以及建筑在自然和当地社区方面所产生的影响进行处理，同时从中受益。学习的最佳方法是去实践！加入到建造的过程中来，创造属于自己的独特空间，是一种非常有效的实践条件。凉亭的设计方案成为一些有益的障碍：航空动力学（通风）、物理学（光扩散）、生物学（光合作用、种植）。这些都有益于指导用户的实践，使居住环境更加环保。

项目名称：BES Pavilion
地点：Ha Tinh, Vietnam
建筑师：Doan Thanh Ha, Tran Ngoc Phuong
项目团队：Chu Kim Thinh
用途：service space for an open community, focusing on the aspects of art and culture
用地面积：234m² 有效楼层面积：123m² 景观面积：111m²
总建筑面积的覆盖率：52.5% 最高高度：5.3m
结构：bamboo, earth, stone 竣工时间：2013.8
摄影师：Courtesy of the architect - p.63^{top}, p.65$^{top\text{-}right, bottom}$, p.66
©Tran Tuan Trung (courtesy of the architect) - p.60, p.61, p.62, p.63^{bottom}, p.64, p.65$^{top\text{-}left}$, p.67

BES Pavilion

BES pavilion is a service space for an open community, focusing on the aspects of art and culture. Located in the central Ha Tinh City, BES (Bamboo + Earth + Stone) is set up from local materials and traditional building methods which are based on the idea of centralizing the users.

The cluster includes various separated spaces which were arranged freely around the center courtyard in order to create numerous views as well as the interaction of light and shadow. Those will help to erase the bound between inside and outside space.

The building's users will have a great chance to approach and to be educated from the functions and effects of the building toward the nature and local community. The best way to learn is to do it! Joining in the building process to create their own specific space is an effective practicing condition. The solutions of the pavilion's design themselves become some useful lesions: aerodynamics (ventilation), physics (light diffusion), biology (photosynthesis, planting). Those will help to direct the users' behaviors in the future for a greener living environment.

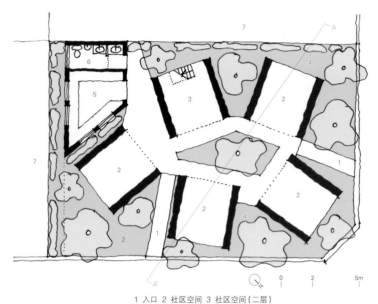

1 入口 2 社区空间 3 社区空间（二层）
4 花园 5 服务区/厨房 6 卫生间 7 现存建筑
1. entrance 2. community space 3. community space (2 levels)
4. garden 5. service area/kitchen 6. restroom 7. existing building
一层 first floor

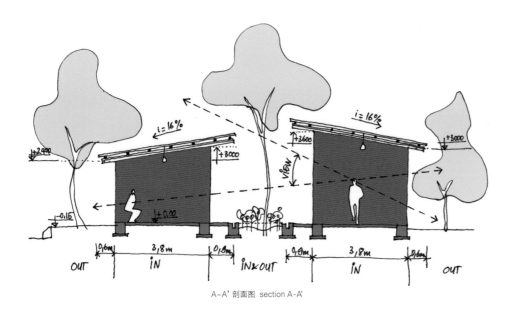

A-A' 剖面图 section A-A'

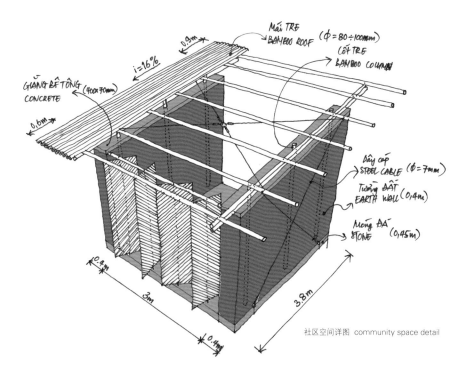

社区空间详图 community space detail

兰溪庭
Archi-Union Architects

兰溪庭位于中国成都国际非物质文化遗产公园内,包括三部分:餐厅、室内庭院以及私人会所。它是一座采用数字制造的语言来诠释中国传统的建筑。

该项目的空间布局展示了对中国南方传统花园的解读。纵向住宅区和庭院的多样化布局展现了传统花园的分层式和多维度空间模式。建筑的屋顶为绵延的山体和翻滚的河流的想象,同时也象征着中国传统斜屋顶文化。

波纹墙的设计来源于对水的数字解读,这是一个灵活的自然理念。最初,当水在时间的维度内展现静止的形态时,这一理念试图抓住这种瞬间形态,且设计改造能对这种瞬间形态提供一种实质的建筑

表达。最后，建筑采用了传统的建筑材料，青砖，以及交错连接技术，但是特殊的空心节点则实现了动态水、光线的视觉效果以及结构墙的透明效果。设计重点为建筑形式以及合理的建造模式。在现在的技术和施工条件中，自动机械不适宜在大规模建筑中铺砖。所有的砌砖都应该手工完成，以和严格的施工规程相匹配。建筑师提供了五种砖缝模板，并且通过排列与分类来形成联合梯度。这样砌砖布局模式便形成了。建筑充分展示了光影交错产生的动态水效果。将数字技术与低技术的建造技术相结合，来实现数字建造，反映了数字技术与当代材料和建造方法的绝佳组合。

Lanxi Curtilage

The Lanxi Curtilage is located at the International Intangible Cultural Heritage Park in Chengdu, China. It is composed of three parts: restaurant, inner courtyard and private club. It is an interpretation of traditional Chinese architecture through the language of digital fabrication methods.

Spatial layout of this project represents a new interpretation on traditional South China garden. Multiple layout of longitudinal residence and courtyard reflects hierarchical and multi-dimensional spatial pattern of traditional gardens. Ups and downs of the building roof embody rolling mountains and rivers, and also func-

项目名称：Lanxi Curtilage
地点：International Intangible Cultural Heritage Park, Chengdu, China
建筑师：Philip F. Yuan
设计团队：Lv Dongxu, Meng Yuan, Alex Han
甲方：Chengdu Qingyang Suburb Construction & Development Co., Ltd.
用地面积：Approx 4,000m² 总建筑面积：3,000m² 有效楼层面积：1,200m²
设计时间：2008.6—2009.3 施工时间：2010.4—2011.10
摄影师：©Shen Zhonghai (courtesy of the architect)

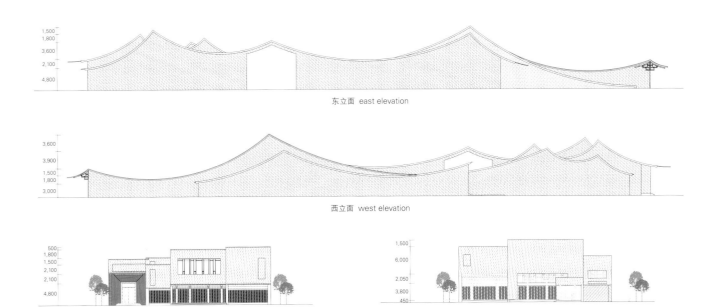

东立面 east elevation

西立面 west elevation

南立面 south elevation

北立面 north elevation

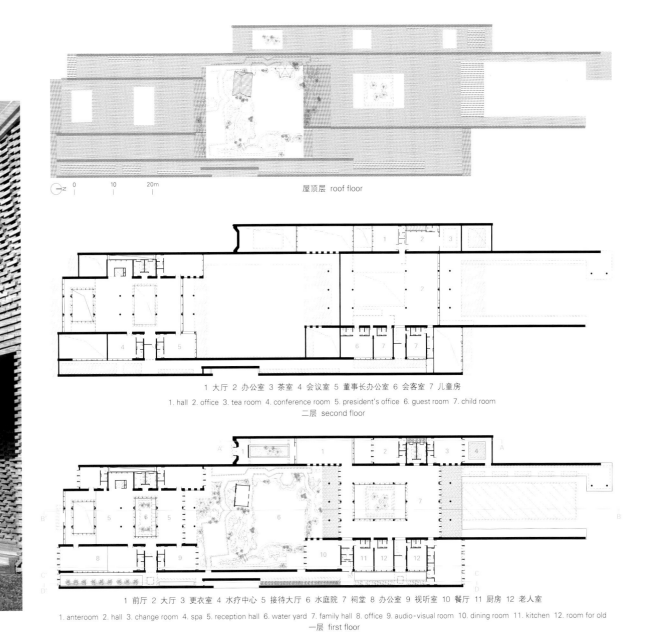

屋顶层 roof floor

1 大厅 2 办公室 3 茶室 4 会议室 5 董事长办公室 6 会客室 7 儿童房
1. hall 2. office 3. tea room 4. conference room 5. president's office 6. guest room 7. child room
二层 second floor

1 前厅 2 大厅 3 更衣室 4 水疗中心 5 接待大厅 6 水庭院 7 祠堂 8 办公室 9 视听室 10 餐厅 11 厨房 12 老人室
1. anteroom 2. hall 3. change room 4. spa 5. reception hall 6. water yard 7. family hall 8. office 9. audio-visual room 10. dining room 11. kitchen 12. room for old
一层 first floor

tion as a metaphor of traditional Chinese sloping roof culture. Design of ripple wall derives from a digital interpretation of water, a flexible natural conception. In beginning, algorithm tries to catch the transient behavior when water tends to be static along time dimension, and design transformation provides literal architectural expression on this transient behavior. Finally, traditional building material, blue brick, and staggering joint process are employed, but special hollow joint process actualizes visual effect of dynamic water as well as light and transparent effect of heavy wall. Design focuses are building pattern as well as reasonable fabrication pattern. Robot is not available in current technical and construction conditions for the accurate expression of bricks with a variety of staggering scales. All brick masonry shall be completed manually and match rigorous construction schedule. Five kinds of blue brick joint template are provided and joint gradient is achieved through the permutation and classification of the five template values. And then masonry pattern is summarized. It is verified that bricks under light and shadow can embody the dynamic effect of water. Combination of the digital design and low-tech fabrication to actualize digital fabrication exactly reflects the combination of digital technologies and local materials and fabrication approach.

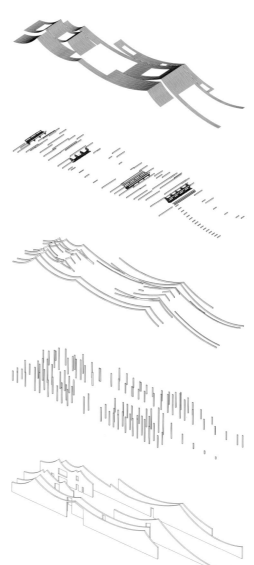

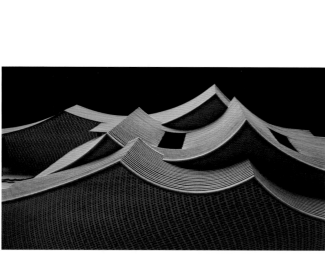

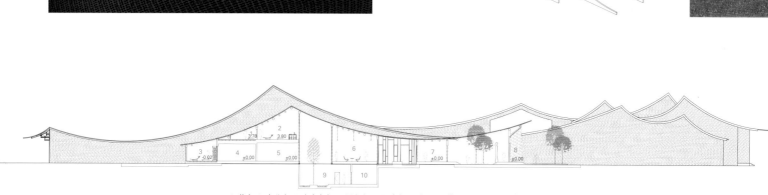

1 茶室 2 办公室 3 水疗中心 4 更衣室 5 运动室 6 大厅 7 前厅 8 主入口 9 庭院 10 机械室
1. tea room 2. office 3. spa 4. change room 5. sport room 6. hall 7. anteroom 8. main entrance 9. courtyard 10. machine room
A-A' 剖面图 section A-A'

1 办公室 2 祠堂 3 接待大厅 4 走廊
1. office 2. family hall 3. reception hall 4. corridor
B-B' 剖面图 section B-B'

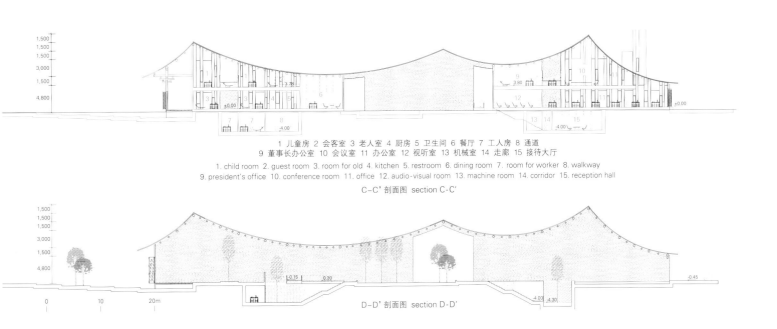

1 儿童房 2 会客室 3 老人室 4 厨房 5 卫生间 6 餐厅 7 工人房 8 通道
9 董事长办公室 10 会议室 11 办公室 12 视听室 13 机械室 14 走廊 15 接待大厅
1. child room 2. guest room 3. room for old 4. kitchen 5. restroom 6. dining room 7. room for worker 8. walkway
9. president's office 10. conference room 11. office 12. audio-visual room 13. machine room 14. corridor 15. reception hall

C-C' 剖面图 section C-C'

D-D' 剖面图 section D-D'

肉桂培训中心
TYIN tegnestue Architects

建筑师将中心的主结构到室内都利用肉桂树的树干来建造，而在门和窗户上应用的工艺的精巧性，是这个项目中给予建筑师印象最深刻的一个方面。

主要结构包括一个大批量生产的Y形支柱，采用螺栓连接至下方的混凝土基础。柱子的布局要服从于楼层平面，而施工体系要确保坚固性和严格性。在面积较大的屋顶表面之下，建筑师建造了五座青砖建筑，一座小型实验室、若干间教室、办公室以及一间厨房位于其间。

在这种规模的项目中，在较短的三个月工期内，后勤将作为一个主要的挑战呈现在眼前。因为七十名工人参与项目其中，八头水牛从森林中运载树木，以及现场实施锯木工作，项目管理变得十分重要。整个项目是由十个简单的细节构成的。基本且务实的设计方法使未接受培训的劳动力来完成这个项目成为可能。

这一地区的建筑所面临的另一项主要挑战是频繁的地震。这个结构已经在几次地震（里氏震级达到五级）中存活下来。以此证明根据不同的工作频率来将不同的建筑构件进行分割的理念是正确的。肉桂合作培训中心已经通过了自然力量的测试。建筑师希望，也坚信，它会完成这一目标，即赋予本地农民以及工人一处安全、卫生且具有社会可持续性的工作场所。

Cassia Coop Training Center

The architects chose to utilize the trunks in everything from the main construction to the interior of the center. The finesse of craftsmanship found on the doors and windows of the center, is some of the most impressive they have witnessed during the projects.

The main construction consists of mass produced Y-pillars, bolted down into a concrete footing. The placement of the pillars subordinates to the floor plans, while the system of the construction secures tightness and rigidity. Underneath the massive roof surface the architects find five brick buildings, and amongst them are a small laboratory, classrooms, offices and a kitchen.

In a project of this size, with a short timeframe of three months, logistics will present itself as one of the major challenges. With seventy workers taking part, eight water buffaloes hauling trees from the forest and an on-site sawmill, project management becomes essential. The entire project is made up of ten simple details. Basic and pragmatic approach to design made it possible to realize this project with an untrained workforce.

Another major challenge of building in this area is the frequent earthquakes. The construction has already survived several quakes reaching over five on the Richters scale. This proves that the idea of separating different building components with different frequency works. Cassia Coop Training Center has passed the test of the forces of nature. The architects hope and believe it will also fulfill its ambition of giving the local farmers and workers a safe, sanitary and socially sustainable workplace. TYIN tegnestue Architects

项目名称：Cassia Coop Training Centre
地点：Sungai Penuh, Kerinchi, Sumatra, Indonesia
建筑师：Gjermund Wibe, Morten Staubo, Therese Jonassen, Kasama Yamtree, Andreas Gjertsen, Yashar Hanstad
学生：Rozita Rahman, Bronwyn Long, Sarah Louati, Zofia Pietrowska, Zifeng Wei
施工：TYIN Tegnestue Architects, local workers
赞助商：LINK Arkitektur
甲方：Cassia Coop
用途：training facility for cinnamon production workers
用地面积：12,000m²
建筑基底总面积：600m²
楼层面积：400m²
造价：EUR 30,000
施工时间：2011.8—11
摄影师：©Pasi Aalto (courtesy of the architect)

1 展览室
2 办公室
3 庭院
4 卫生间
5 保安室
6 教室
7 实验室
8 厨房

1. showroom
2. office
3. courtyard
4. toilet
5. guard house
6. classroom
7. laboratory
8. kitchen

一层 first floor

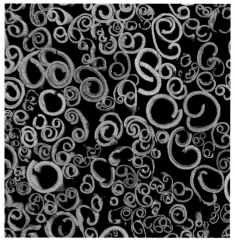

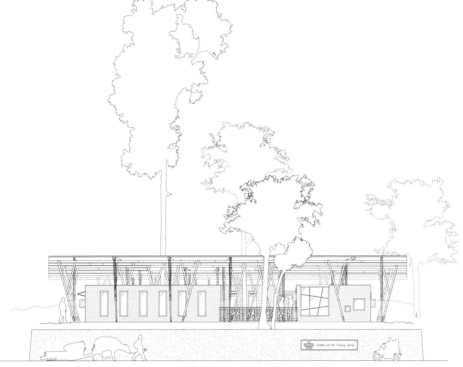

北立面 north elevation

西立面 west elevation

A-A' 剖面图 section A-A'

0 2 5m
B-B' 剖面图 section B-B'

乡土情怀
VERNACULAR FICTION

现在许多国家，甚至包括经济增长缓慢的国家，不断地将资金投入到场地的重塑中，以吸引更多的游客，提高国家GDP。

之所以这样做，正是呼应了人们想要逃避现实的固有需求，这是我们的社会特征，尤其对于消费至上主义者来说，当代紧张的生活节奏往往更加强化了他们逃避现实的愿望。而这些度假胜地通常会带给游客一种不切实际的幻想，并试图通过奢侈的理念和传统氛围让他们觉得自己备受日常工作职责的宠爱和呵护。经过几十年持续不断大规模的景观开发和对历史建筑胡乱的乡土定位，人们最终展示出了新态度，开始负责任地规划和设计旅游市场。

现在的"乡村小屋""度假别墅"和精品酒店再次对"本土化"一词的真正含义提出了质疑。人们通过一些虚构的体验，来设计这些旅游性建筑的布局，达到为客人提供逃避现实的庇护所的目的。这些布局试图将游客置身于某种原始的生活状态，让他们觉得能够回归自然，无需为日常生活分心，能够远离日常商业活动的烦扰。所以相对于建筑本身的功能来说，它们的人文气息更浓。

上述案例表明当再次建造本土化建筑时，人们会对自然的真正价值有了新的认识。人们会借助当地的材料和技术，仅仅加入必要的运作材料。

有些人会说这和主题公园的休闲方式如出一辙，但是最终，这不比那些出售快餐和全包服务的特许景点要好吗？

Nowadays many countries (even those with very poor economy growth) keep investing into a "restyling" process, in order to attract more tourists and improve their GDPs.
They do that by appealing on the innate need of escapism that characterizes our society, always more consumerist stressed by the pressing rhythms of contemporary life. These resorts often transport the tourist into a fake dream, where the perception of luxury and a classical touch try to give him the feeling of being spoiled and protected by his daily duties.
After decades of this continuous trend of massive landscape exploitations and abuse of vernacular references to historical architecture, the tourism market is finally witnessing a new responsible approach to design and planning.
Today's "Rural houses", "Albergo Diffuso" and Boutique Hotels are re-questioning the true meaning of the word, "vernacular". These tourism topologies, while preserving their intention to provide escapism for their guests, do it by means of fictional experience. They try to transport the tourist into a sort of original state, where he can live with the primary essence of the place without the distractions and commodities of everyday life. They become then anthropological rather than architectural projects.
These examples show a renovated awareness of the true value of nature, when dealing with the Vernacular, and they do that by finding materials and technologies on site, adding exclusively what's necessary for its activation.
Some could say that this might be as well the recreation of a theme park, but ultimately, isn't it better than franchised resorts with fast food and all inclusive formulas?

If one sector of the construction industry might be considered safe from the catastrophic fall of Western economies, it would have to be tourism and leisure. Tourism has experienced growth almost inversely proportional to the Western economical nosedive. If on one hand we have experienced growing unemployment and plummeting internal GDPs, on the other, who doesn't jump nowadays on a low cost flight to reach the closest holiday place for a short "detoxification" weekend? The need for escape – escapism – is growing, and as never before, the reason for this tendency is strictly embedded in the root of the word itself: escape!
Hedonism and entertainment have experienced exponential growth through the centuries, starting with rural retreats and wellness centers such as the Roman thermae, and culminating in today's tourism capitals (Las Vegas, Macau, Dubai, Benidorm: the list could span the globe). Health, culture and entertainment have intermingled, evolving along with our cultures and societies.
Entertainment locations have also followed an exponential "iconization" of the concept of luxury; the actual physical appearance of the resort has developed a globalized language which has gradually mixed the original intentions of providing entertainment and relaxation with the complementary rationales of business and commerce. This mixture has occurred everywhere in the same way. In Dubai as in Delhi, resorts can be surrounded by slums – it doesn't matter – as long as the formula for fictional hedonism continues to be respected. All providers offer the same occidental franchised spaces, where the same lounge music plays low while elegant ladies sip their martinis.
Combining profit motive and a "classical touch", a growing proliferation of leisure destinations seek to camouflage their global formats with a layer of local "identity". In fact, however, the reference to vernacular archetypes is reduced to mere visual power while, deprived of any actual architectural and functional basis.
The era of modern icons starts with the great example of Las Vegas, where the city was transformed from a three-dimensional to a two-dimensional process. Like a stage, the entire city has become an oversized scenario, a sequence of facades, presenting a mes-

9 Spa/a21 Studio
Basiliani Hotel in the Rock Dwellings of Italy/Domenico Fiore
Hornitos Hotel/Gonzalo Mardones Viviani
Fasano Boa Vista Hotel/Isay Weinfeld

Vernacular Fiction/Maurizio Scarciglia

If there is a branch of the building industry that perhaps has been spared the meltdown of the Western economy, it may be tourism and leisure. Against the collapse of Western economies, tourism is instead growing. If, on one hand, we are living scenarios of growing unemployment and plummeting internal GDPs, on the other, today who would not jump to a cheap flight to the nearest resort for a weekend of "detoxification"? The desire to escape – escape from reality – is stronger than ever, for reasons deeply connected to the roots of a world that is globally escaping.

The leisure industry has exponentially developed over centuries, starting from the rural retreats and health centres such as the Roman Thermae, reaching its peak today with the tourism metropolis (Las Vegas, Macao, Dubai, Benidorm: tourist cities are a global phenomenon). Wellbeing, culture and entertainment merge and evolve with our culture and society.

Entertainment venues follow the concept of luxury; the actual physicality of resorts has developed into a common language, gradually combining the original purpose of providing entertainment and leisure with the complementary logic of commercial exchanges. This combination occurs in the same way throughout the world. In Dubai, as in Delhi, resorts are surrounded by shantytowns – it does not matter – as long as this fictional hedonistic mode continues to be accepted. All resorts offer the same Western franchises where, without exception, the quiet music of time-killing plays, and elegant ladies sip martinis.

Leisure destinations, combining profit purposes and "traditional experiences", have become increasingly widespread and seek a local "identity" to cover their generic models. Yet, in fact, the identification of vernacular archetypes stops simply at the visual aspect, losing the substantial architectural and practical foundation.

Las Vegas set the paradigm for the contemporary era. In this city, the urban process has changed from three-dimensional to flat. Like a stage set, the whole city is one giant stage scenery, a series of facades, transmitting the mes-

sage of luxury and happiness to an entire social class, the tourist. The background has become a functional apparatus whose sole purpose is to support the mega urban facade: a kind of "back office", blown up to the city scale.
Resorts, hotels, and sometimes entire cities have become "vernacular monsters", making anachronistic references to the past. The apparently compulsive need for those vernacular icons is evidence of the power of classicism on the consumer imagination.
In the chimera of contemporary resorts, landscape types are transformed into visual types. However, the memory of ethereal atmospheres, of relaxing environments, is enough to deceive the consumer: The result is a scenography, a trompe l'oeil urbanism. The dream and exotic, luxurious appearances offer small consolation to the oblivious tourist of the 21st century. For a few weeks per year, these theme parks and resorts become a time machine for tourists with a limited budget and desire for comfort.
With everything from casinos to shopping malls and entertainment centres, from restaurant chains to indoor ski complexes, contemporary resorts have been transformed into modern cities, whose only exotic quality is their "Truman Show" setting[1].
The contemporary "visual culture" has made everything look familiar. Contemporary tourists are looking for familiarity: they want to feel at home in a strange place. This has led to concentrated tourist infrastructures and mega structure complexes (hotel + apartments + mall + cinema + expo + anything), which are clustered together. In this sense, architecture and landscape are part of a single system, characterized by stratification and controlled spatial experience[2].
Fortunately, there are emerging resort typologies which are, by definition and design, sustainable and more respectful of the true meaning of "vernacular". "Rural houses" (Agriturismo), the Albergo Diffuso (Diffused Hotel) – both Italian formulas – and boutique hotels and many more new approaches are re-introducing the value of a genuine experience of a place. This experience starts from the recreation of fictional realities, including life in a former time when no electricity, heating, wifi or other commodities were on offer. This imaginary, almost surreal experience becomes even more appealing to the tourist, eager as never before to forget his or her daily duties and dive into a dream, even if only for a few days.
The striking aspect of these typologies is their pursuit of a subtle, strategic approach to the promotion of urban quality and economic sustainability by minimizing new interventions. In the diffused hotel, for example, the typical corridor that accesses each room is replaced by city streets, which experience a new life, having one more reason to be maintained, cleaned, and decorated. No need exists for new construction, only the simple reactivation of the existing.
These examples show a renovated awareness, among designers and developers, of the fundamental values of "nature, identity and uniqueness", materialized through the use of local technologies and materials and by respecting the local typologies. These are the only values that contribute to glue architecture with the context and that ultimately recreate a real dream of escape for the tourist. In this sense, the vernacular becomes the answer to satisfy the desire of local and the escape from everyday's busyness.
Even if sustainable, these formulas, Las Vegas-like, recreate fictive settings that resurrect past traditions, serving the Las Vegas pur-

位于拉斯维加斯的Venetian酒店是现今典型商业综合建筑的典范之一

The Venetian Hotel in Las Vegas as one of the examples of today's typical commercial complex

从娱乐城到大型购物中心及娱乐中心,从连锁餐饮店到室内滑雪场,当代的度假胜地已经变革为现代都市,唯一醒目的特征就在于它们的"楚门的故事"背景不同。[1]

当代的"视觉文化"使得所有的事物看上去都似曾相识。当代的游客寻找的就是这种亲切感:他们要的是在陌生的环境中也能感到自在。这随之产生了集合型的旅游基础设施和大型结构建筑群(旅馆+公寓+商场+电影院+展览会+其他),它们全都集中在一起。从这种意义上来说,建筑和景观隶属同一体系,这个体系具有层次化的特点,且空间受限。[2]

幸运的是,新的旅游胜地类型正在形成,而且,从定位和设计来看,具有持续性,对"乡土"一词的真正含义表现得更为尊重。"乡村小屋"(餐饮酒店)、Diffused酒店——两者均为意大利模式——以及精品酒店,还有更多新的模式正在重新提出真实体验一个地方的新价值。这种体验始于体验虚构生活的游戏,包括体验以前那种没有电、没有供暖、没有局域网和其他商品的生活。这种虚构的、几乎超现实的体验对游客更具吸引力。他们比以往任何时候都更急于想要抛却他们的日常杂务,躲进一场梦里,哪怕是几天的时间也好。

这些模式显著的方面就是它们通过尽量减少干预,来追求巧妙地、有策略性地提升城市品质和经济持续性。例如,在Diffused酒店,城市街道取代了通往各个房间的特设走廊,为游客提供一种全新的生活体验,也为其得到养护、打扫、装饰增强了必要性。这类酒店无需进行新的建造,只需简单地重新利用既存设施。

这些例子表明设计者和开发者对"自然、个性、唯一"这些基础价值有了重新的认知,具体体现在对当地技术和材料的运用及对当地模式的尊重上。唯有这些价值能够粘合建筑与环境,最终为游客再造逃避现实的美梦。从这种意义上说,乡土既能满足当地需求,又能帮助人们逃离外界琐事。

即使是可持续性的,这些模式,如同拉斯维加斯,再造虚构的环境,以重现过去的传统风俗,只是为拉斯维加斯式的梦想服务。所以,虚幻的享乐主义仍然从根本上控制着旅游市场。

9水疗中心_a 21 studio

"9水疗中心"由9个酒店组设而成,内设一间小酒吧和一间餐厅,提供水疗、泥浴和矿物浴。水疗中心的有趣之处在于它的建筑构造和管理结构,它的可持续性在于运用当地材料和人工:工程雇佣了当地九十名泥瓦匠、木匠和手工艺人,花费九个月的时间建造而成。建筑材料结合了就地采掘的干叠石、木料和当地的椰叶。

屋顶构造采用了传统技术。屋顶为三层,覆盖在木质框架的上面:分别为一层20mm厚的木镶板、一层防水隔膜和一层30mm厚的椰叶。从附近建筑中收集来的古老家具、门、桌子、板凳和带图案的瓦片赋予了水疗中心独特的乡土风情。

小酒吧的位置被抬高,既连接了内外空间,又为顾客观景提供了新的视角。正如建筑师所言,自然是核心价值,自然之美从这座建筑

pose of serving up a dream. Therefore, fictive hedonism remains the sovereign germ of the tourism market.

9 Spa_a21 studio

9 Spa is a set of nine hotel houses offering a spa, mud and mineral baths together with a small bar and restaurant. The interesting aspect of this project lies in its construction and management structure which are sustainable because it is based on the use of local materials and working power: Ninety village masons, carpenters and craftspeople were enlisted to build the hotel in 9 months. The project uses a combination of dry-stacked stone quarried on the site, wood structure, and local coconut leaves.

The roof structure adopts a traditional technique. Above its wooden structure, the roof has three layers: 20mm-thick wood panelling, a water-proof membrane and a 30mm-thick layer of coconut leaves. Old furniture, doors, tables, chairs and patterned tiles gleaned from nearby buildings give the building a distinctive vernacular look.

The small bar is elevated, linking the outer and interior spaces and offering customers a new viewpoint. As the architects say, nature is treated as the core value; its beauty can be contemplated from every corner of the project.

Basiliani Hotel in the Rock Dwellings of Italy_ Domenico Fiore

At the world heritage site of Sassi di Matera, in southern Italy, the prehistoric homes are carved into the stone escarpment, and a curious blend of stunning views of the Gravina Canyon District and the Archaeological Park defines a unique historical landscape. Basiliani Hotel occupies a renovated chain of houses, half carved out of the ground and half built of local tufa stone. The hotel covers three levels, respecting the original grade of the site, with a restored exterior that allows multiple points of access through the original doors. From the exterior, the project's scope is invisible to the naked eye. The stone masonry and barrel vaults are covered with a layer of white paint that reflects an abundance of natural light into the cavernous spaces while preserving original forms and textures.

This lodge, like several similar examples in Southern Italy, represents a strategic perpetuation of the existing legacy while offering the comforts of a contemporary resort.

Hornitos Hotel_ Gonzalo Mardones Viviani

The hotel in Hornitos sits on a plateau overlooking a promontory, 32 meters above sea level in the Atacama Desert in northern Chile. Because the site is a pristine, immaculate place, the main chal-

Basiliani酒店坐落于意大利南部被称为世界遗产遗址的圣堤马特拉,这座酒店刻铸于石坡中
Basiliani Hotel, located at the world heritage site of Sassi di Matera in southern Italy, carved into the stone escarpment

9水疗中心,是基于当地材料和人工建造起来的
9 spa, based on the use of local materials and working power

意大利岩屋的Basiliani酒店_Domenico Fiore

Basiliani酒店坐落于意大利南部被称为世界遗产遗址的圣堤马特拉,这座史前小屋刻铸于石坡中,格雷韦纳峡谷区令人惊奇的风景与考古公园的奇妙结合确立了这里独特的历史景观。

Basiliani酒店拥有一组翻修的小屋,这些小屋都属半地下结构,房屋的另一半由当地泉华石建成。考虑到此地的原有坡度,酒店分三层,通过翻修的外部结构,人们可以从不同方向进入原门。从外部看,建筑规模用肉眼是看不到的。砌石和桶形穹窿上涂了一层白漆,可以将充足的自然光反射进房间的深处,同时又可保留原有的结构和质地。

这些乡村小屋,如同意大利南部其他类似的小屋,既是现存遗产继续巧妙保留下去的见证,同时又增强了当代旅游胜地的舒适度。

Hornitos酒店_Gonzalo Mardines Viviani

Hornitos酒店坐落在智利北部阿塔卡马沙漠的高原上,高出海平面32m,可以俯视整个海岬。由于此地属于原始的未经破坏的区域,所以在此建酒店最大的考验就是如何将生态破坏降到最低。半地下的建筑布局将区域进行了大小划分。大的区域用来建造酒店,一个个散落的小型体量则给了游客体验广袤沙漠的机会。每个体量都由钢筋混凝土建成,表层粉刷成沙漠的颜色,以增强拟态效果。

酒店占地5800m²,拥有38个房间、18个小木屋、一个室外泳池、一个室内温水泳池、一个礼堂、多间会议室、多间餐厅、多间休息室、多间健身房等等。

强烈的日光通过一组连续的过渡区来调节,日光由遮阳区逐渐向非遮阳区过渡。公用区大多呈开放式,可以自然通风。大楼利用屋檐和格子架来遮挡落日的照射。屋顶被设为休闲区,以便游客眺望大海。酒店安装了海水处理装置,可以从海中直接取水。

Fasano Boa Vista酒店_Isay Weinfeld

Fasano Boa Vista酒店位于巴西的波尔图菲利斯港,距离圣保罗100km,占地面积为750公顷,集住宿与食宿为一体。除了酒店,还包括私人别墅、一间水疗中心、一个儿童俱乐部、一个马术中心、一个体育中心、一座爱畜动物园、两个18洞的高尔夫球场、一个高尔夫俱乐部、一个游泳池和一片占地面积为242公顷的森林。

酒店大楼为巨型水平结构,两侧为对称的侧楼。侧楼中间的主楼用来承办接待、招待、办公和内务整理活动。

住房排列在接待区左右两侧的侧楼。每个侧楼由13个立体模块构成,共容纳39间客房,在立面上以石板和砌砖加了边框以做区别。长长的走廊沐浴在柔和的自然光中。地下酒吧和餐厅一直延伸到户外的大平台,平台伸出湖面。

建筑材料选用木、石、灰泥、天然纤维和皮革,营造出简朴的氛围,优雅、素净、朴实无华,立刻让人联想到20世纪五六十年代圣保罗的乡村度假胜地。

lenge was to intervene with as little invasiveness as possible. The semi-buried architectural layout presents a larger volume containing the hotel and scattered smaller volumes to allow visitors to experience the vastness of the desert. Each volume is made of reinforced concrete pigmented with the colors of the desert, intensifying the intention of mimesis.

The 5,800m² hotel has 38 rooms, 18 cabins, an outdoor swimming pool, an indoor heated swimming pool, an auditorium, meeting rooms, dining rooms, lounges, gyms, and so on.

The strong sunshine is controlled via a sequence of intermediate spaces that allow a gradual transfer from protected spaces to those exposed to the sun. Common circulation spaces are mostly open, allowing natural ventilation. The building protects itself from the setting sun with eaves and lattices. The roof is treated as a relaxation space for contemplating the sea. Equipped with a saltwater treatment plant, the complex takes its water directly from the sea.

Fasano Boa Vista Hotel_Isay Weinfeld

Fazenda Boa Vista is a residential and hospitality complex located on a 750-hectare property in Porto Feliz, Brazil, 100km from São Paulo. In addition to the hotel, it is comprised of private villas, a spa, a kids club, an equestrian center, a sports center, a petting zoo, two 18-hole golf courses, a golf clubhouse, a swimming pool and 242 hectares of woods.

The hotel building is defined by a large horizontal structure with two symmetrical wings, flanking a core body that hosts the reception, entertaining, office and housekeeping facilities.

All accommodations lie along the left and right wings emanating from the reception area. Each wing consists of a sequence of thirteen cubic modules housing all 39 guestrooms, on the facade marked by frames enclosing slabs and brickwork. The long hallways are bathed in soft natural light. On the underground level, the bar and restaurant extend outdoors onto a large deck projecting over the lake.

The choice of materials – wood, stone, stucco, natural fibres and leather – contributes to an understated mood, at once elegant, plain and unpretentious, reminiscent of the Sao Paulo countryside resorts of the 1950s and 1960s. Maurizio Scarciglia

1. *The Truman Show*, Peter Weir, Paramount Pictures, USA, 1998. In The *Truman Show*, Truman Burbank lives his entire life trapped on a TV set built to broadcast his life as a reality show for the real world.
2. George Katodrytis, "The Dubai Experiment: 1 Accelerated Urbanism", *Al Manakh*, Moutamarat, AMO, Archis, 2007

9水疗中心

a21 Studio

"9水疗中心"由9个酒店组设而成，内设一间小酒吧和一间餐厅，提供水疗、泥浴和矿物浴。这座建筑坐落在通往岩石山体的途中设置的平台处，是一处褶皱场地，使酒店隐藏在下坡重要的区域内。

建筑师采用了本土的建造技术和材料，迎合了当地的风俗习惯来对建筑进行管理。他们雇佣了九十名本村泥瓦匠、木匠以及工匠，以在九个月的工期内建造这座酒店。项目设计是将干叠石材与木结构和椰树叶相结合，并且在场地右侧进行挖掘。

这个中心是建在不同的角度背景中，通过一段距离（被认为是每座房子都拥有的入口大厅）来相互隔离，不仅仅允许雨水自由地从山顶流淌下来，还实现了下方区域的通风。此外，屋顶结构利用榫接头技术，采用传统的方式来建造。屋顶位于木结构之上，为三层，20mm厚的木板赋予天花板以优美的外观，同时分别将所有梁柱、防水薄膜以及30mm厚的椰树叶层连接起来。此外，项目还充分利用附近建筑内的旧家具，如，门、桌子、椅子以及花纹瓦片，赋予建筑独特的外观，以及随着时间的增长旧事物所产生的美和宁静。

酒吧的座位少于12个，酒吧置在高于地面的地方，因此将室内外

空间连接起来,并且为顾客提供了一个全新的观角,同时还未与现存的自然风景相接触。换句话说,通过采用必要的方法,自然成为核心的价值,它的美可以在项目的任何角落里得到展现。

总之,9座酒店由一个连续的木屋顶来相互连接,反映了其周围的环境和景观。通过利用如岩石和木材的当地材料以及旧家具,9水疗中心赋予现存项目一种非凡的价值。

9 Spa

9 spa is a set of nine hotel houses with spas, mud and mineral baths together with a small bar and restaurant. The buildings are perched in the folds of halfway terrace up to a rock hill, which makes the hotels hideaway from the eventful area downhill. The architects use indigenous building techniques and materials, and adopt local custom to manage the project. 90 village masons, carpenters and craft persons were enlisted to build the hotels in a period of 9 months. The project was designed as a combination of dry-stacked stone with wood structure and coconut leaves, quar-

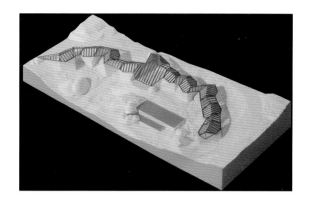

ried right on the site.

The houses are set up in different specific angles and placed separately by a distance, considered as an entrance lobby for each house, to not only let rain water run down easily from top of the mountain but also allow ventilation to the below area. Moreover, the roof structure is constructed in a traditional way by mortise and tenon joint techniques. Above this wooden structure, the roof is a combination of 3 layers, 20mm thick wood panel, which gives an aesthetic look to ceiling and links all beams together, water proof membrane and 30mm coconut leaves, respectively. Besides, the project also makes use of old furniture from nearby buildings such as doors, tables and chairs or patterned tiles, giving the buildings a distinctive look, the beauty or serenity of old items that comes with age.

The bar, with less than a dozen seats, is above from the ground, thereby linking the outside space to the interior and offering a new viewpoint to the customers, while not touching the existing nature. In other words, by any means necessary, nature is treated as the core value, that its beauty can be contemplated at every corner of the project.

In conclusion, nine hotel houses are linked by a continuous wooden roof, reflecting its surround environment and landscape. By using local materials as rock and wood together with old furniture, 9 spa gives an extraordinary value to the existing project.

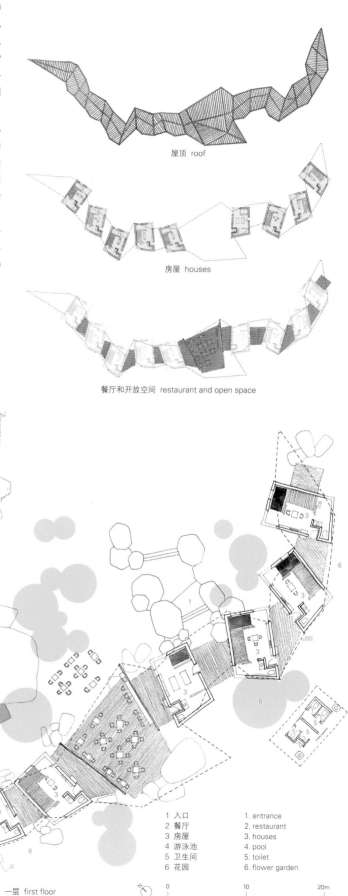

屋顶 roof

房屋 houses

餐厅和开放空间 restaurant and open space

1 入口　　1. entrance
2 餐厅　　2. restaurant
3 房屋　　3. houses
4 游泳池　4. pool
5 卫生间　5. toilet
6 花园　　6. flower garden

一层 first floor

0　　10　　20m

1 入口 2 餐厅
1. entrance 2. restaurant
北立面 north elevation

A-A' 剖面图 section A-A'

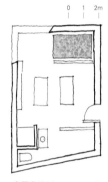

房屋类型1 house type 1

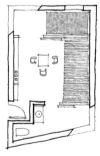

房屋类型2 house type 2

项目名称：9 spa
地点：Nha Trang, Vietnam
建筑师：a21 Studio
甲方：I-resort
用地面积：1,080m²
总建筑面积：450m²
材料：rock, wood, coconut leaf, used furniture and tiles
竣工时间：2013
摄影师：
©Hiroyuki Oki (courtesy of the architect)-p.92~93, p.94, p.95, p.96, p.97
©Nam Phan (courtesy of the architect)-p.91 (except as noted)

意大利岩屋的Basiliani酒店
Domenico Fiore

Gravina Canyon

Gravina Canyon

Basiliani大酒店是一家设计酒店,以现代的手法展现了传统窑洞式的小型结构,使其现代化与众多世界范围内的一个超群背景中的历史产生对比。

酒店在2010年进行了全面的修复。它地处Sasso Caveoso的中心,即Sassi区域内最古老的地方,位于格雷韦纳峡谷区域的边界处。酒店由10间客房组成,采用现代极简主义风格来进行分别设计,以基线为特色,来保护原始形式和结构,包括钟形的水池(当地人用来存储雨水),这个系统与其他系统一起使Sassi区成为最古老的也是最佳的建筑仿生学案例之一。白色是墙体的主打颜色,黑色则主导和引导着地面的颜色,发挥着测量和绘制的作用,木质家具的颜色则为红色。

设计意在唤醒一处维度空间,在这处空间内,自然与设计共存,以强化人类和环境之间的情感共鸣。人们能够从房间内看见格雷韦纳峡谷区域令人窒息的史前窑洞景观,以及考古公园(包括联合国教科文组织的场地)。房间是由"tufo"(一种当地的柔软的石头)制成,一些房间是建造的,而其他房间则是挖掘形成的。

项目名称:Basiliani Hotel in the Rock Dwellings of Italy
地点:Matera, Italy
建筑师:Domenico Fiore
建造商:Artedil srl
有效楼层面积:900m²
设计时间:2005 竣工时间:2010
摄影师:©Piermario Ruggeri Milano(courtesy of Basiliani Hotel)

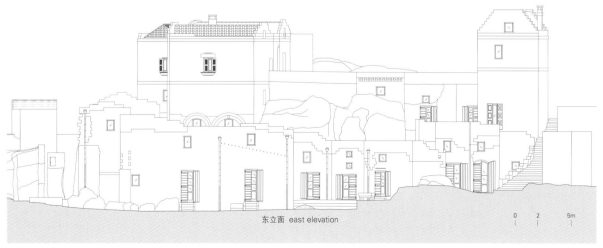
东立面 east elevation

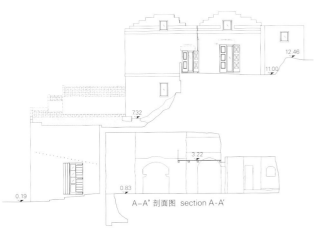

A-A' 剖面图 section A-A'

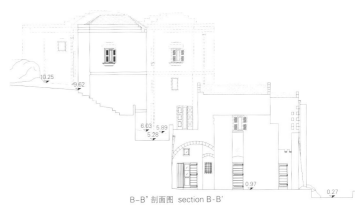

B-B' 剖面图 section B-B'

first floor

Basiliani Hotel in the Rock Dwellings of Italy

Basiliani Hotel is a design hotel which elaborates the minimal structure of the traditional cave-house in a contemporary perspective, contrasting modernity with the past in one of the most extraordinary world settings.

The hotel was completely restored in 2010. It is located in the heart of Sasso Caveoso, the oldest part of the Sassi area, on the edge of the Gravina Canyon District. It is made up of 10 guest rooms that have been individually designed in a contemporary minimalist style, characterized by basic lines, preserving the original forms and structure, including the bell-shaped cisterns that natives have used for the collection of rain water (a system that has concurred to make the Sassi one of the oldest and best preserved example of bio-architecture). White is the dominant color of the walls, the dark floor invades and leads, measures and draws each space, and red is the color of the wood furniture.

The design is intended to evoke a dimension where nature and design coexist, enhancing the emotional synergy between human and the environment. From the rooms people can grasp breathtaking views of the prehistoric caves in the Gravina Canyon District and the Archeological Park (Unesco sites). The rooms are made in "tufo", a local tender stone, some are built, others are caved.

乡土情怀 Vernacular Fiction

Hornitos酒店
Gonzalo Mardones Viviani

Hornitos酒店的业主为Caja de Compensación Los Andes，坐落在智利北部阿塔卡马沙漠的高原上，高出海平面32m，可以俯视整个海岬。因为这里是一处特殊且完美的区域，因此项目面临的主要挑战便是嵌入场地，尽量不产生入侵的感觉。因此，建筑师采用了半地下的建筑布局，在水平方向设置一个大型体量，用来容纳酒店，小型体量（小屋）则散布在场地周围，以覆盖当前区域，与广袤的沙漠融为一体。每个体量都是由钢筋混凝土构成，着以沙漠的颜色，以突出拟态效果，使人们经过时忽略建筑的存在。

酒店面积为5800m²，拥有38个房间，有36个房间分布在18间小屋中，此外，酒店还设有一座室外泳池、一座室内恒温游泳池、礼堂、会议室、餐厅、休息室以及健身房等等。

来自于阿塔卡马沙漠的强烈阳光对人们来说十分合适，由一系列过渡空间来调节，这些空间产生了遮阳区逐渐向非遮阳区的过渡。酒店大部分常规的交通流线以及空间都是开放式的，来产生空气流，允许空气持久地循环。建筑利用屋檐以及格子架来遮挡阳光的照射。屋顶是一处休闲区域，一处可以休息和看海以及观远景的平台，也是一处将其本身置在通风室（在屋顶产生通风，使其远离影响酒店室内空气条件的阳光直射）的平台。这座综合体直接利用海水，内设一个盐水处理设备，能够净化废水，且将剩余的盐沉入到酒店入口附近的室内泳池中。

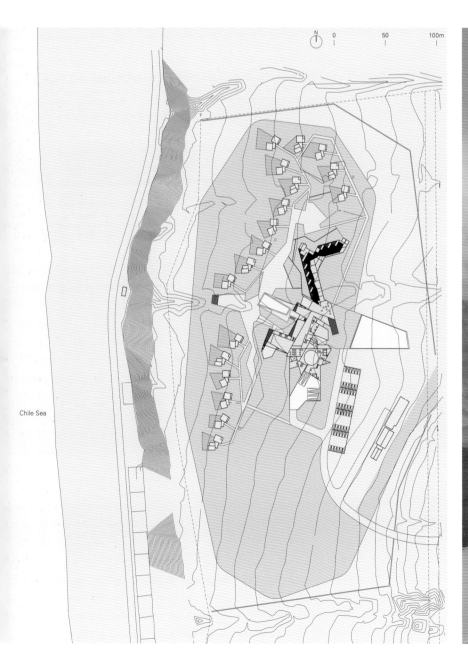

Chile Sea

西北立面 north-west elevation

东北立面 north-east elevation

0 10 30m

Hornitos Hotel

The hotel for the Caja de Compensación Los Andes in Hornitos is located on a plateau over a promontory 32 meters above sea level in the Atacama Desert, North of Chile. As this is a privileged and immaculate place, the main challenge was to intervene trying to be the less invasive possible. Therefore, the architects opted for a semi-buried architectural layout carried out horizontally with a larger volume containing the hotel, and smaller volumes scattered on the land(cabins), in order to cover the immediate area and join the vastness of the desert. Every volume has been made of reinforced concrete pigmented with the colors of the desert intensifying the intention of mimesis to try to pass unnoticed.

The hotel has 5,800m² and 38 rooms(36 in 18 cabins), an outdoor swimming pool, an indoor heated-swimming pool, an auditorium, meeting rooms, dining rooms, lounges, gyms, etc..

The strong sunshine, proper from the Atacama Desert, is controlled by a sequence of intermediate spaces that allow a gradual transfer from the protected places to the ones exposed to the sun. Common circulations and spaces of the hotel are mostly open, generating air currents that allow constant air circulation. The building protects itself from the sun setting with eaves and lattices. The roof, treated as a relaxation place, a terrace to stay and watch the sea and the distant landscape, is a great deck creating under itself an air chamber that allows to ventilate the roof and keep it isolated from the direct sunshine affecting the adequate air conditioning in the internal spaces of the hotel. The complex is provided with water directly from the sea, having a salt water treatment plant inside the facilities, which cleans water to be used and deposits remaining salts into an indoor pool next to the hotel access.

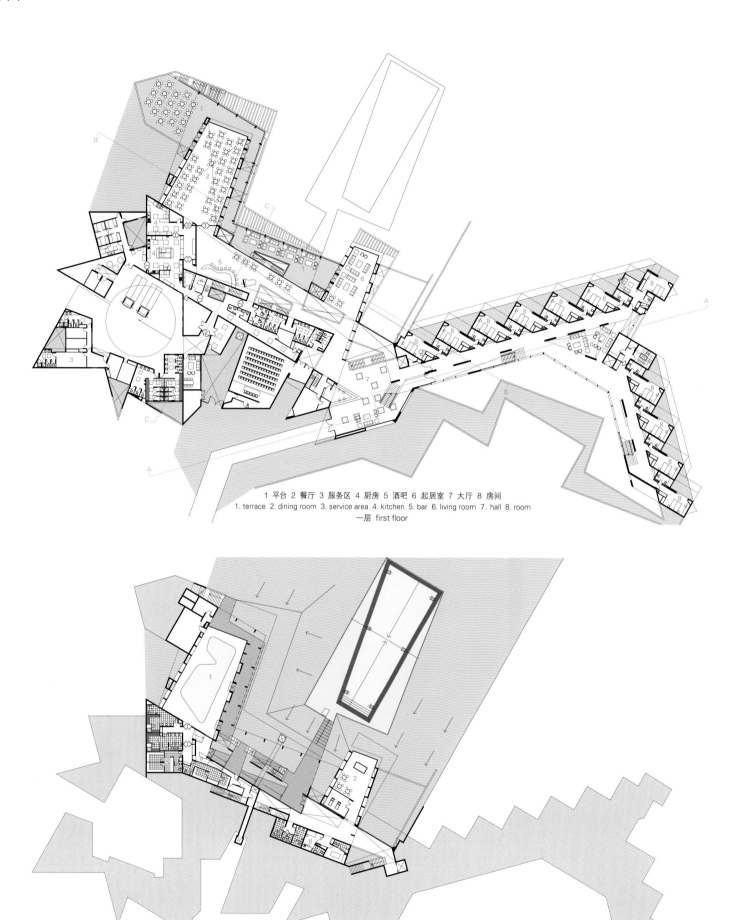

1 平台 2 餐厅 3 服务区 4 厨房 5 酒吧 6 起居室 7 大厅 8 房间
1. terrace 2. dining room 3. service area 4. kitchen 5. bar 6. living room 7. hall 8. room
一层 first floor

1 室内泳池 2 健身房 3 桑拿室 4 浴室 5 卫生间 6 办公室
1. inner pool 2. gym 3. sauna 4. bathroom 5. restrooms 6. office
地下一层 first floor below ground

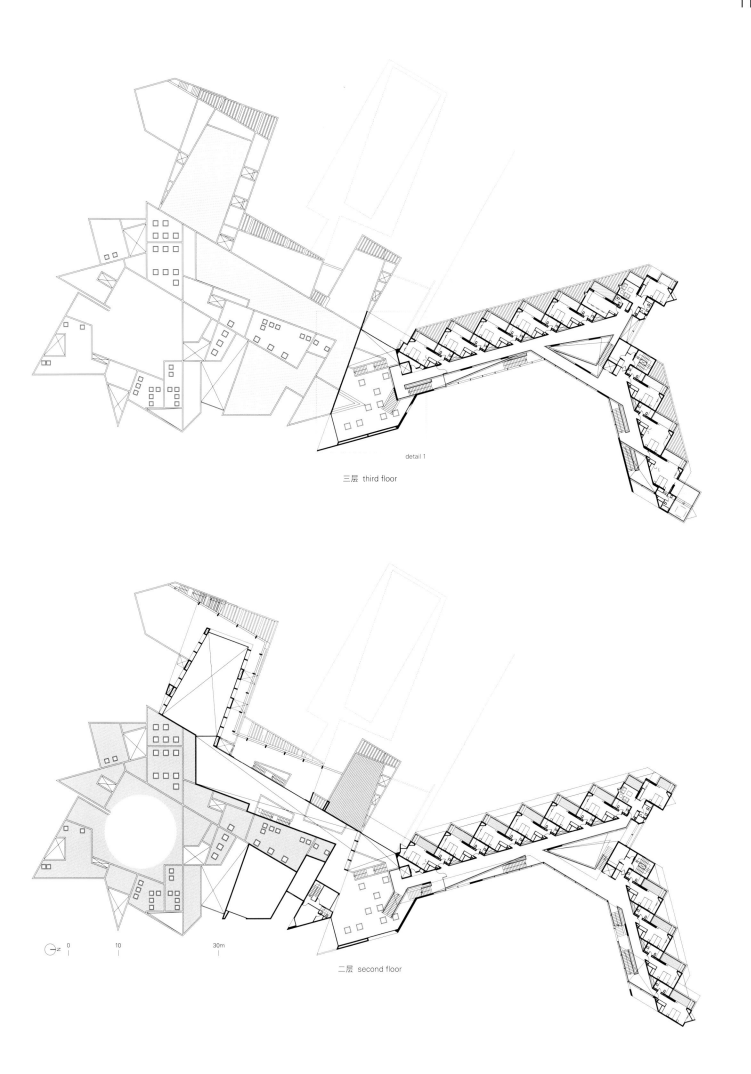

detail 1

三层 third floor

二层 second floor

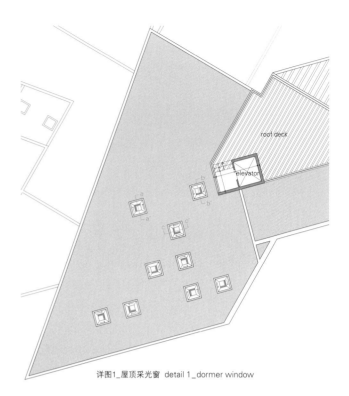

详图1_屋顶采光窗 detail 1_dormer window

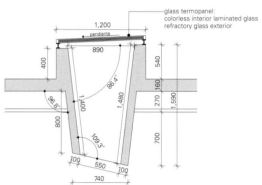

a-a' 剖面图 section a-a'

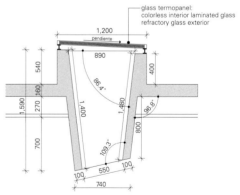

b-b' 剖面图 section b-b'

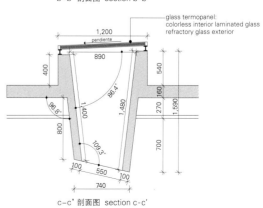

c-c' 剖面图 section c-c'

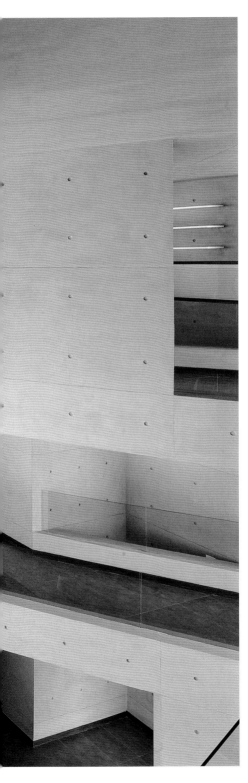

项目名称：Hornitos Hotel
地点：Hornitos, II Region, Chile
建筑师：Gonzalo Mardones Viviani
项目团队：Gonzalo Mardones Falcone, María Jesús Mardones Falcone, Luis Morales Gatto, Claudio Quezada Fuentes, Emilio Ursic Marechau, Manuel Fuentes Ramos, Alberto Reeves Droguett, Claudio Leiva Brito, Cristián Romero Valente, Claudio Carrasco Fuentes, Francisco Valdés Donoso, Alessandro Beggiao
工程师：Alfonso Larraín 照明工程师：Paulina Sir
景观建筑师：Cecilia Rencoret
业主：Caja de Compensación Los Andes
用途：38 rooms, swimming pool, auditorium, meeting rooms, dining rooms, lounges, gyms, etc.
总建筑面积：5,800m² 竣工时间：2012
摄影师：©Nico Saieh (courtesy of the architect)

1 健身房 2 室内泳池 3 卫生间 4 平台 5 起居室 6 餐厅
1. gym 2. inner pool 3. restroom 4. terrace 5. living room 6. dining room
B-B' 剖面图 section B-B'

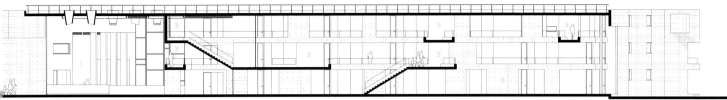

1 大厅 2 起居室 3 走廊
1. hall 2. living room 3. corridor
A-A' 剖面图 section A-A'

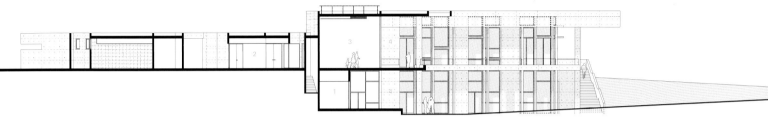

1 卫生间 2 走廊 3 酒吧 4 平台
1. restroom 2. corridor 3. bar 4. terrace
C-C' 剖面图 section C-C'

乡土情怀 **Vernacular Fiction**

Fasano Boa Vista酒店
Isay Weinfeld

1 主入口大门	1. main entrance gate
2 直升机场	2. helipad
3 马术中心	3. equestrian center
4 马术中心会所	4. equestrian center clubhouse
5 马球别墅1	5. polo villas-1
6 马球别墅2	6. polo villas-2
7 儿童俱乐部	7. kids club
8 Fasano别墅	8. Fasano villas
9 高尔夫俱乐部会所	9. golf clubhouse
10 Fasano酒店	10. Fasano hotel
11 水疗中心	11. spa
12 高尔夫球场	12. golf course
13 入口大门	13. entrance gate

西南立面 south-west elevation

东北立面 north-east elevation

Fasano Boa Vista酒店位于巴西的波尔图菲利斯港，距离圣保罗100km，其所在的酒店综合体占地面积为750公顷。除了酒店，还包括私人别墅、水疗中心、儿童俱乐部、马术中心、体育中心、爱畜动物园、两个18洞的高尔夫球场、高尔夫俱乐部、游泳池和一片占地面积为242公顷的森林，森林中设有很多的湖泊。

酒店建筑设置在场地的最高点之一，俯视着其中的一座湖泊，且人们还可在此观看日落。这是一座大型水平结构，两侧为对称的侧楼——它们绘制出了光变曲线，一条为凹曲线，另外一条则是凸面线，分布于主楼的两侧。主楼用来承办接待、招待、办公和内务整理活动。

主楼：在建筑的中心横轴右侧，入口小径引导人们到达接待大厅，这条小径位于木质绿廊之下，贯穿一座郁郁葱葱的花园。人们从接待处向前行，便能看见休息室与阳台连为一体，逐步地展现出更为宽阔的视野，最后，将湖泊与更广阔的绿色景观毫无保留地呈现在人们面前。在地下一层，酒吧与餐厅延伸至室外的一个大型平台中，平台伸出湖面，以用于冥想和游泳。

侧楼：住房排列在接待区左右两侧的侧楼——一侧设有26间标准间，另一侧则设有11间复式套房、一间双床复式套房以及一间残疾人客房。每侧侧楼由13个连续的立体模块构成，共容纳39间客房，在立面模块上以围合石板和砌砖的边框来做区别。通向各个房间的长长的走廊沐浴在柔和的自然光中，光线通过一系列沿着东北立面设置的预制混凝土板来进行过滤。

建筑内充满了低调的修饰和舒适氛围。建筑材料选用木、石、灰泥、天然纤维和皮革，这些材料与家具共同营造出简朴的氛围，优雅、素净、朴实无华，立刻让人联想到20世纪五六十年代圣保罗的乡村度假胜地。

Fasano Boa Vista Hotel

Fasano Boa Vista Hotel is located in a 750-hectare hospitality complex in Porto Feliz, 100km away from the city of São Paulo, in Brazil. Besides the hotel, it comprises yet private villas, spa, kids club, equestrian center, sports center, petting zoo, two 18-hole golf courses, golf club, swimming pool and 242-hectare woods punctuated with innumerous lakes.

Placed on one of the highest spots of the property and overlooking one of the lakes and the sunset, the hotel building is defined by a large structure of pronounced horizontality, composed by two symmetric wings – they draw light curves, one slightly concave and the other slightly convex – flanking the core body (reception, entertaining, office and housekeeping facilities).

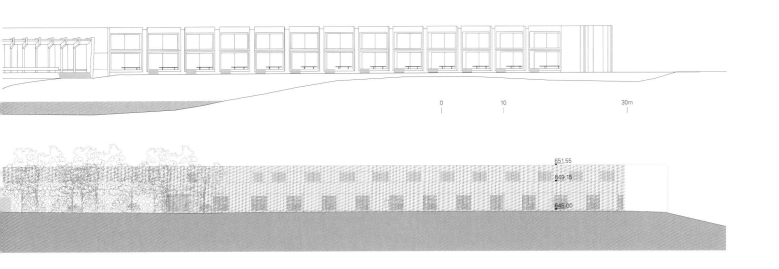

Core body: right on the central transverse axis of the building, the entrance pathway leading to the reception hall is set under a wooden pergola and through a lush garden. From the reception forward, lobby and veranda develop in succession, integrated and gradually allowing for wider perspectives until, ultimately, unveiling the lake and the extensive green landscape. On the underground level, the bar and restaurant extend outdoors onto a large deck projecting over the lake, which serves for contemplation as well as swimming.

Wings: to the left and right of the reception, all accommodations are to find – 26 standard rooms to one side, and 11 duplex suites+1 duplex two-bedroom suite+1 room equipped for guests with disabilities to the other. Each of the wings is made up of a sequence of thirteen cubic modules housing all 39 guest rooms, clearly marked on the facade by the frames enclosing slabs and brickwork. The long hallways leading to the rooms are bathed in soft natural light, filtered by a sequence of precast concrete slats standing all along the northeast facade.

The ambiance all over is that of low-key eloquence and coziness. The choice of materials – wood, stone, stucco, natural fibers and leather – as well as the furnishings, all contribute to this understated mood, at once elegant, plain and unpretentious, reminding of those resorts that existed in the 1950s and 1960s in the São Paulo countryside.

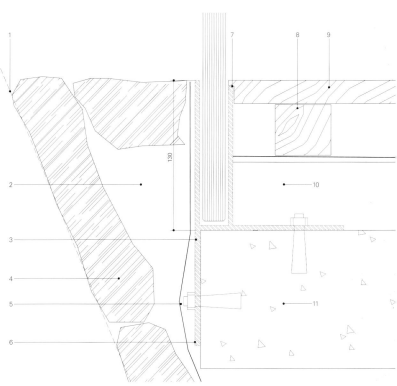

1. imaginary alignment line with incline, lateral walls made of stone and tops of upper beam made of wood
2. provide for metal lath to settle ground 3. metal section aligned to end of slab
4. raw stone cladding 5. provide for waterproofing 6. stainless steel angle bracket fastened parabolt
7. stainless steel structure to support railing 8. joist 9. wooden deck 10. subfloor 11. slab

平台饰面剖面详图 deck finish section detail

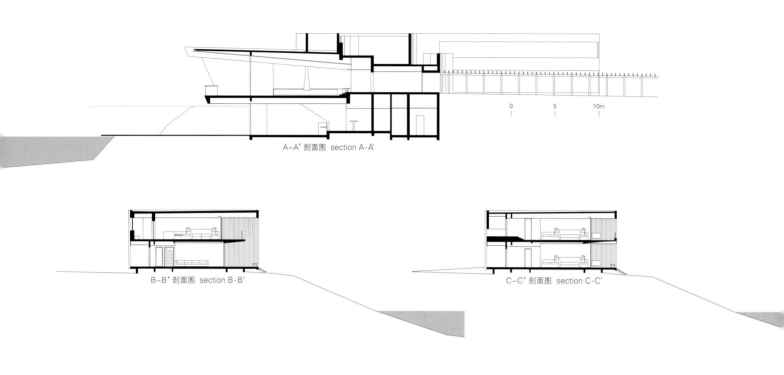

A-A' 剖面图 section A-A'

B-B' 剖面图 section B-B'

C-C' 剖面图 section C-C'

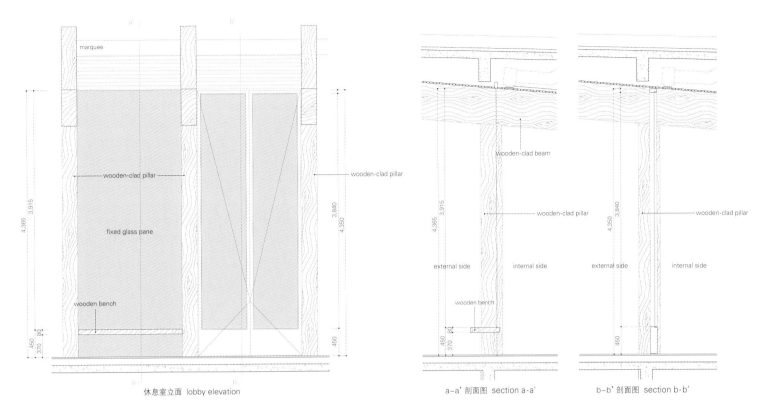

休息室立面 lobby elevation   a-a' 剖面图 section a-a'   b-b' 剖面图 section b-b'

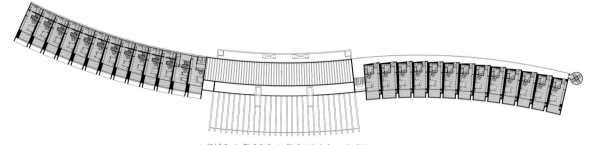

1 机械室 2 复式套房 3 复式双床套房 4 标准间
1. mechanical room 2. duplex suite 3. duplex 2-bedroom suite 4. standard room
三层 third floor

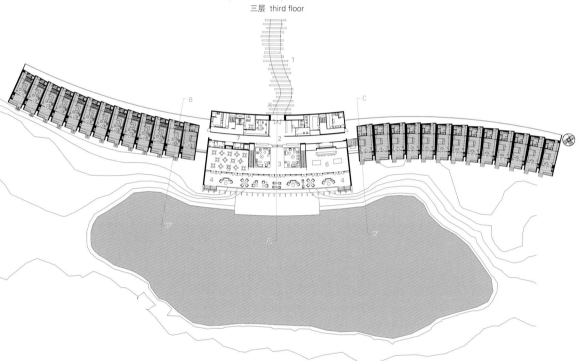

1 主入口 2 接待处 3 大厅 4 阳台 5 游戏房 6 斯诺克酒吧 7 办公室 8 复式套房
9 复式双床套房 10 残疾人客房 11 标准间
1. main entrance 2. reception 3. lobby 4. veranda 5. game room 6. snooker bar 7. offices 8. duplex suite
9. duplex 2-bedroom suite 10. room equipped for guests with disabilities 11. standard room
二层 second floor

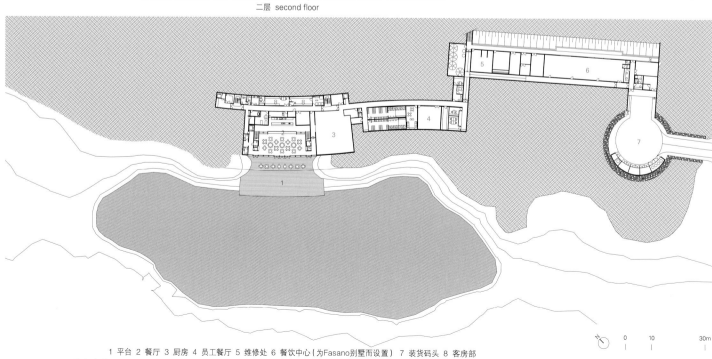

1 平台 2 餐厅 3 厨房 4 员工餐厅 5 维修处 6 餐饮中心（为Fasano别墅而设置） 7 装货码头 8 客房部
1. deck 2. restaurant 3. kitchen 4. staff canteen 5. maintenance 6. catering (for the villas Fasano) 7. loading dock 8. housekeeping
一层 first floor

复式套房侧楼 duplex suites wing

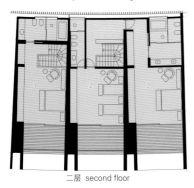

二层 second floor

一层 first floor
1 复式套房 2 复式双床套房 3 残疾人客房
1. duplex suite 2. duplex 2-bedroom suite
3. room equipped for guests with disabilities

标准间侧楼 standard rooms wing

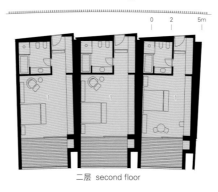

二层 second floor

一层 first floor
1 标准间
1. standard room

项目名称：Fasano Boa Vista Hotel
地点：Porto Feliz, São Paulo, Brazil
建筑师：Isay Weinfeld
合作商：Domingos Pascali, Marcelo Alvarenga
项目经理：Monica Cappa
项目团队：Juliana Garcia, Wellington Diogo, Juliana Scalizi, Guilherme Leme
总建筑商：JHS-F
结构工程师：Benedictis Engenharia Ltda
基础工程师：Apoio
电气、机械以及水管工程师：
Grau Engenharia De Instalações Ltda
自动化：Bettoni Automação e Segurança Ltda
空气调节：Thermoplan Engenharia Térmica Ltda
景观设计：Maria João
总建筑面积：8,651m² 有效楼层面积：6,911m²
竣工时间：2011. 8
摄影师：©FG+SG Architectural Photography

金孝晚
HyoMan Kim

传统景观

文化认同
通过对传统与现代并存的建筑的全球化类型的研究,我期待韩国文化扎根于现代生活。
我试图探索新的建筑基因逐渐演变的传统。

漫步
韩国传统建筑的内外部空间都有着有趣而神奇的特点:设计方案多样;门庭的开合断断续续地演变;空间的彼此渗透和循环,这使得斜坡地貌的上下空间更为富足,向人们展示着风景如画的洞口。所有这些都有引人漫步其间的吸引力。
"漫步"已成为我作品中最重要的空间概念,它是感性空间的可持续方案,使情感生活空间最大化。因此在我所有的跃层空间设计项目中都用到了它。
Lim GeoDang是我采用此设计的第一个项目。

内向性
MaDang(韩式内部庭院)是一个缩影,空间柔和恬静。它既保证了个人隐私性,也预防了犯罪,可以用来安排日常生活的所有活动。对于GaOnJai项目,生态花园采用了"MaDang"以保护隐私并预防犯罪;而对于LimGeoDang和KyeangDoklai项目来说,采用"MaDang"是为了使各类花园拥有最大限度的地域;对于WaSuJai项目,则是为了与韩式旧屋共存;而Island House设有MaDang是为了远离公众视野,并与邻近河流和谐共存;对于NoksungHun和BuYeoDang来说,我们将"MaDang"作为外部空间,成为私人花园。

曲线
传统韩国屋顶的三维曲线结构是高昂活力的标志性特征。它与周边的自然地貌和谐一致,而且它的形象显著提升了空间活力。另一方面,它与现代社会城市景观形成对比。对于GaOnJai项目来说,我们引入曲线

Tradition Scaping

Cultural Identity
Through the research on glocal type of architecture which is traditional and modern at the same time, I have expected that the identity of Korean culture takes root on our modern life.
I try to explore the evolved tradition for our new architectural gene.

Strolling
Inner and outer spaces of traditional architecture of Korea have interesting, dramatic character of various programs, intermittent evolvement of opening and closing, penetration and circulation of spaces, producing rich spaces of up and down of sloped topography, introducing picturesque opening. They all have the attribute of "strolling" of interest.
"Strolling" has been the most important spatial concept of my work and it is a sustainable program of space in sensibility to maximize the life of emotional space, so it has applied to all my projects with skipfloor system of space.
LimGeoDang was the first project I gave its shape to.

Introversion
"MaDang"(inner court of Korea) is one's own "microcosm", with warm and static character of space. It keeps the individual privacy and prevents crime. It is an empty place to accommodate infinite programs of daily life. For GaOnJai, "MaDang" was adopted for ecological garden to keep privacy and prevent crime. For LimGeoDang and KyeongDokJai, to get the maximum area of various gardens. For WaSunJai, to coexist with old-Korean house. For Island House, to avoid from public eyes and to be in playful harmony with the adjoining river. For NokSungHun and BuYeonDang, to keep private garden, we applied "MaDang" as exterior space.

Curving
3-dimensional curved structure of traditional Korean house roof is the iconic character of soaring dynamism. It is harmonized with the surrounding context of natural topography of Korea and its image typically promotes spatial dynamism. On the other hand, it is constrasted with urban landscape of modern society. For GaOnJai, we introduced this vocabulary of curving to harmonize with the surrounding mountains, to invest the spatial dynamism

GaOnJai
KyeongDokJai
办公园

传统景观 / 金孝晚

GaOnJai
KyeongDokJai
Office Park

Tradition Scaping/HyoMan Kim

这一词汇，使之与环绕的山峦相称，激发所有空间的流通活力，在房展会中不同寻常的村落中植入文化特征。KyeongDokai项目采用曲线，来给城镇角落注入活力，营造建筑文化。而Silverboat采用曲线，同样给城市角落注入活力，利用有弧度的象征手法，使人们想起河流、微风和帆船。

隔离

传统的房间与房间之间不设通道的分隔式布局有一些劣势。其中之一就是即使在雨天或是雪天，人们也必须经过室外空间。但另一方面，在经过时，我们可以享受自然。如果我们在非日行通道的设计上也借鉴这一创意，就能创造出有趣的空间设计，让我们与自然协调一致。GaOnJai的办公室和LimGeoDang的书房采用的就是这种自然体验的设计方案。

漂浮与移动

体现在"ru（楼）"结构中的浮动性和厚层悬臂屋顶的动态浮动性是韩国传统建筑的典型特征。GaOnJai、LimGeoDang、HakIkJai、Light House中的"Ru"式底层架空结构不仅实现了有趣的空间渗透，设有覆顶的室外空间，还能体现现代生活的都市活力，这种浮动性也是内部空间活力展示的产物。

画框

传统建筑墙面上镶有美景如画的洞口能从视觉上提升人们漫步其间的兴趣。这些边框就像博物馆里的艺术品。对于WaSuJai项目来说，这种设计是认可旧式韩国居舍的存在；对与NokSungHun项目来说，是把美景框入墙上。

现代生活的根基是传统

传统应当是现代生活文化精髓的体现。我们试图开创我们这一时代的多种混合文化。

of circulation of all the spaces and to insert the cultural identity in monumental village of house expo. For KyeongDokJai, it is used to give dynamism on the corner of the town and to shape the "architectural nature". For Silverboat, to give vitality at the corner of city and to remind the river, wind and sail, the typology of curving was applied.

Separation
Traditional layout of separating masses without corridor has some disadvantages. One of them, for instances, is that one must pass through the outer space even in the rain or snow. But on the other hand, we can enjoy the nature during the walk. If we adapt this idea to our nondaily pass, we will be able to invent interesting programs of the space that will frequently put us along with the nature.
Office in GaOnJai and study room in LimGeoDang were designed with this course of experience of nature.

Floating & Moving
Floatability that is expressed in the structure of "Ru(楼)" and dynamic floating of the deep cantilevered roof are typical characters of traditional architecture of Korea.
"Ru"-like piloti structures of GaOnJai, LimGeoDang, HakIkJai, Light House, not only have got interesting space of penetration, covered exterior spaces, but also have urban vitality in modern life. This typology of floatability has been the result from the expression of dynamism of inner space.

Pictorial Frame
Picturesque landscape framed openings in the walls of traditional architecture are a visual factor to promote the interest of "strolling" of spaces and the frame looks like a piece of art in the museum. For WaSunJai, it is used to recognize the being of the old Korean house. For NokSungHun, to frame the landscape on the wall.

The root of modern life is tradition.
Tradition should be the essence of cultural identity of modern life. We try to explore hybrid gene of culture of our time. HyoMan Kim

嵌入的非线性场景的冲击效应

TaeCheol Kim + HyoMan Kim

TaeCheol Kim GaOnJai项目的设计似乎是创造多样外部空间的一种尝试。将房屋最大限度地贴近边界线，以使庭院更大一些。正如您过去尝试设计的房屋一样，这体现了建筑理念从外部空间到内部空间的发展。而KyeongDokJai项目则正好相反，采用的是面朝开阔的北面的漂浮庭院。我想，这和拥有类似规模的SoDaHun项目也有所不同。和SoDaHun项目相比，在GaOnJai项目的设计上您做了哪些改动呢？

金孝晚 由房屋的主体部分围绕的GaOnJai项目和KyeongDokJai项目的内部庭院是生机勃勃的，与之不同的是，由栅栏和大量房屋组成的SoDahun项目的内部庭院由于场地狭小而不那么有生气。GaOnJai项目的"匚"形内部庭院三面被房屋围合，北面是半开放式的，由一片小竹林和对面住宅区隔离开，以此遮挡不雅观的景色同时保持私密性。

因此，GaOnJai项目的内部庭院令人满意。相反，SoDaHun项目的内部庭院是"┓"形户外空间。两边是低栅栏及其上层的不锈钢网组成的半开放式围墙，另两边则由屋群环绕。

由于它的极端封闭性，我们试图在内部庭院的院门上安装透亮的有机玻璃和透明玻璃来缓解这种封闭性。

KyeongDokJai项目的内院是"H"形户外空间，两侧由房屋围绕，南面朝向上层的邻里住宅，以利于采光。但是北面的道路通过半透明的铝屏来半开放地朝向公共绿地。

和SoDaHun项目相比，KyeongDokJai项目的内部庭院更开放且更有活力，拥有开放的自然景观和多样的空间特色。

TCK GaOnJai项目和KyeongDokJai项目的门廊都不在一楼而在二楼，我们是不是可以认为这是在有意增加可视性？这种可视性对住户在漫步时可能会造成不便。有对此不满意的客户向您抱怨过吗？

HMK 我认为一个人的房子就是他自己的缩影。这所房子的主门就是和现实世界的分界线。

从主门到入门的过程是一种缓冲，也是一种介绍。我们期待住户因为逃离了现实社会而享受到日常生活的一丝喜悦。著名评论家JongGeun Lee教授把这处能带着愉悦的心情闲逛的地方称为"Yu（游）"，我就引用他的"Yu"来解释这一"漫步"过程。

The Shock Effect of Inserted Nonlinear Scenes

TaeCheol Kim GaOnJai seems to be an attempt to create a variety of outside space by putting the mass at maximum closeness to the boundary line so as to secure a rather larger courtyard. This shows how the architectural ideas develop from the outer area to the inner area as your designs of residential houses used to try. In contrast, the floating courtyards face the open north in Kyeong-DokJai. I think this makes difference with the similar-scaled SoDa-Hun. Comparing to SoDaHun, what kind of differences did you try in GaOnJai?

HyoMan Kim Unlike the active innercourts of GaOnJai and KyeongDokJai which are surrounded with main masses of house, the innercourt of SoDaHun which is composed of fence and a mass of house is passive, because of its extreme small area of site. "匚" shaped innercourt of GaOnJai is closed by masses of house in three side and the north side of site is semi-opened against opposite residential block by small bamboo forest, to avoid unfavourable landscape and to keep privacy.

Consequently, the innercourt of GaOnJai is satisfactory. On the other hand, the innercourt of SoDaHun is "┓" shaped outdoor space, two sides of which are semi-opened by lower fence and higher stainless mesh and the other two sides are surrounded with a body of house.

Due to its extreme closure, we tried to mitigate the closure by inserting translucent polycarbonate glass and transparent glass on the gatedoor which is located in the innercourt.

The innercourt of KyengDokJai is a "H" shaped outdoor space, two sides of which are surrounded with the body of the house and the other side of south is opened to upper side of neighbor houses to introduce sun light, but the northside of front road is semi-opened to public green through the translucent aluminium screen.

In comparison with SoDaHun, the innercourt of KyeongDokJai is a more open, active, extrovert space which has open landscape and various spatial characters.

TCK GaOnJai as well as KyeongDokJai has its porches not on the 1st floor, but on the 2nd floor. Can we say this was intended as "visibility"? The visibility provided when strolling might cause inconveniences for the residents. Have you ever been complained by your clients unsatisfied?

HMK I think one's house is a microcosm of his own and the main gate of the house is the boundary between real world.

The space of the process from main gate to entrance door is a kind of buffer and introduction. We expect inhabitant to enjoy a sense of delight in daily life, because of escaping from their real society.

GaOnJai项目（左）和KyeongDokJai项目（右），当人们漫步其间时，能够感受到空间和视觉方面的舒适性

GaOnJai(left) and KyeongDokJai(right); spatial and visual amenity is given while strolling.

在许多小房子案例中，花园的大窗户总是拉着窗帘以防被他人看到，门前的路和小花园交汇是不多见的。我们把GaOnJai项目和KyeongDokJai项目的主门设在高出道路一层半的位置，目的是在道路和主要的起居空间之间保持隐私性。

因此要到达一层半高的主门就得有一定的距离，我们试图把这种拥有视觉小插曲的愉悦漫步空间介绍给大家。

设计建筑物是一场永无止境的选择和放弃的游戏。我认为，选择漫步，就应该放弃短距离的便捷性。我们应当考虑到不便距离的局限性。但在总体房屋上，多上几步台阶也并没有多么的不方便，这只是业主的选择问题。

对于我的建筑作品，业主评价颇高，因为他们在屋内漫步时能感受到尊重，同时也会感觉屋内空间比一般空间要大。

TCK 我期待从您的作品中看到最现代的外观设计，用金属材料表现出的直观的对角线和曲线。但是您使用的凹曲线属于韩式风格。我认为，韩国传统建筑不像中国和日本建筑那样相对直接和具有象征性，它们更倾向于与周围环境之间有种隐喻和和谐的关系。您能说说您是怎样看待您的建筑作品中包含的韩国式元素的吗？

HMK 毫无条件的对民族主义的尊重性。韩国元素是体现我的作品特征的有趣要素之一。随着时间的流逝，因为厌恶，我十分抵触全球化和统一化。

凹曲线是古代韩国乃至东方都使用的典型建筑语言，但是在现代生活中，由于对西方建筑语言的耳熟能详，人们反而不熟悉凹曲线了。在一些因为空间和概念原因而需要凹曲线的设计上，我会用到凹曲线。从20世纪90年代的Namgang建筑到近期的GaOnJai项目和KeyongDokJai项目，我们因为适当的目的而采用这种有特色的凹线。最重要的是，凹曲线不只是外形的一个方面，也是提升活力的空间因素，因为凹曲线是半阴阳体。

TCK 在KyeongDokJai项目中，就其作为住宅而论，剧场性的空间是非常独特的，我认为只有把这个场景描绘给每天生活在这里的居民之后，这种设计才有可能实现。居住在此的居民能在这处空间创造出各种氛围，您对这个剧场式空间设计有什么期待呢？

HMK KyeongDokJai家的所有成员在日常生活中都喜欢弹钢

Famous critic, professor JongGeun Lee named this space of promenade with enjoyment "Yu(游)" and I am explaining with "strolling" to quote his "Yu".

In many cases of small house, to avoid the problem that large windows of garden are always closed with curtain because of being seen each other, it's unusual interaction between front road and small garden, we located the entrances of GaOnJai and KyeongDokJai at 1.5 floor higher level than front road for keeping privacy between the road and main living spaces.

Therefore, to reach the entrance at 1.5 floor high, there needs a physical distance, we tried to introduce the space of "strolling" with enjoyment of spatial and visual incident.

To design the architecture is a game of endless choice and abandonment, I think, to choose the interest of "strolling", you should abandon convenience of short distance. We should consider the limit of inconvenient distance, but in general house, it's not inconvenient distance about several steps. It's just a problem of choice of the owner.

As for my works of built-up, the owners have favorable review because they have felt dignity on the strolling inside of their house and felt larger than normal state.

TCK What I expect from your work is the modernist appearance with direct and firm diagonal lines and curves using the metal material. However, the concave curve you use belongs to the Korean style. In my opinion, Korean traditional architecture tends to have a metaphorical and harmonious relationship with the environment unlike Chinese and Japanese architecture whose style is comparatively direct and indicative. Could you tell me what you think about the element of Korean style in your architectural work?

HMK In no position to respect nationalism, Koreaness is one of the interested objects as motive to set the identity of my works.
As time goes by, I became a resistant against the globalization, unification because of sick.
The "concave curve" was a characteristic language of architecture of the East and ancient Korea, but it's an unfamiliar shape in modern life, because of the familiarity with western language of architecture. I have used this term, "concave curve" in the projects which need its properties for contextual, conceptual reasons. From Namgang Building in 1990s, to recent projects of GaOnJai, KyeongDokJai, we have introduced this characteristic curve for proper purpose.
Most important thing is it's not only a factor of shape but also the spatial factor to promote the dynamism, because it is a kind of hermaphroditism.

GaOnJai项目，U形的建筑环绕着庭院
GaOnJai. U-shaped building embosoms the courtyard.

WaSunJai项目，庭院被建筑和假墙所环绕
WaSunJai. The courtyard is surrounded by the buildings and pseudo walls.

琴、弹吉他、敲鼓。

把兴趣爱好诠释在空间设计的结果便是音乐厅式的房屋的出现。倾斜的内置小山就是为满足其他设计需求而得出的多重解决方案。

它是一个斜式通道，以把阳光从屋顶空间引入下部的客厅。它还通过跃层垂直连接主卧、吧台、书房和儿童房，成为一个庞大的、包含所有公共生活空间的多元化空间，这也是这所房子的主要起居空间中最具特色的地方。在这个内置小山中，因为书房、吧台、客厅、餐厅的相互连接，在这儿就能看到日常生活中的各种活动。

我们把生活中活跃的场景想象为家庭友好音乐会或是有客人的联欢会。这处空间的所有地方都是舞台，同时也是普通的观众席。

TCK KyeongDokJai项目的北面的铝屏就是一个隔开邻里视线的屏障。然而我认为这个设计的主要意图在于创造一种对比度强的造型。这也是您一直尝试的前所未有的形式。虽然功能在建筑中非常重要，但是您能不顾其他功能目的而专注于造型吗？比方说，斜列式书架因为是倾斜的可能会非常不方便。您怎么看？

HMK 首先，铝屏的作用是保持私密性而不受外界干扰，并隔离噪音和阳光。

城市要求的鲜明轮廓即是在城市角落有流畅优美的形式。这就使得屏障要有曲线，这个有弧度的屏障因为我对韩式风格的追求而转换为凹形屏障。因此所有这些要求都保证了铝屏的存在。

我认为这种形式是内部空间组织及环境要求的体现。在我几乎所有的作品中，内部和外部空间都是由跃层结构连接在一起的。因此，它很自然地塑造出具有活力的体块形式。如果大家对内部空间没有概念，就会认为外形仅为外形而存在，但我认为没有创新的空间就不会有创新的外形设计，因为外形存在的原因是为了空间。如果结果是好的，动机的转换就不重要了。

要保留建筑的外形，它就应当存在的必要性。一旦它失去存在的理由，就应当被改造或摧毁。因为房屋的功能是为了生活，房屋外形是生活的一种形式，没有功能的房屋，形式是毫无生命的。

TCK In KyeongDokJai, the theatrical space is a very peculiar element considering that it is in a residential house. I think this design can be possible only when you can portray the scenes with the residents using and living in the space every day. What was your expectation of this theatrical space being used and staged by the residents who would create a variety of atmospheres in it?

HMK All the members of family of KyeongDokJai enjoy to play guitar, piano, drum in common life.

The result of translation of such character of hobby into spatial program was a concert hall-like space, and this sloped, interior hill was a multi-solution for the other requirements of design of this house.

It is a sloped tunnel to pass the sunlight from upside of this space to downside living space and it's a vertical route to interconnect master bedroom, bar, study, child bedrooms by skip floor system, and it's one large, multi-space contains all the public living spaces. It is a characteristic place of major living space of this house. In this interior hill, divers behaviors are produced in common life because of interacting between study room, bar, living room, dining room.

We imagined active scene of life like friendly concert of family or carnival concert with guests. All the places of the space are stage, at the same time they are all general seats.

TCK The aluminum screen on the northern side of KyeongDokJai functions as a covering shield from the neighboring eyes. However, I think the primary intention of this design is to create an intense formativeness since you have always tried such unprecedented forms. Although function is very important in architecture, can the formativeness be treated as itself regardless of the other functional purposes? For example, the diagonal bookshelves might be very inconvenient because it is slanted. What do you think about this?

HMK Requirements of aluminium screen are, first of all, keeping privacy from invasion of outer passenger's eyes, and filtration of noise and sunlight.

Urban requirement for the edge of city that is to be fluid form in the corner of the city, made this screen curved and this curved screen was transformed into concave screen by my pursuit of Koreaness. Consequently, such all that requirements sustain the existence of aluminium screen.

I think, the form is the result of organization of inner space and contextural requirement. In almost all my works, inner and outer spaces are organized with skip floor system, and so, naturally, it shapes a dynamic form of masses. If one have no information for inner space, one think that the shape is made only for the shape, but I think there cannot be innovative shape without being innovative spaces, because the reason of shape is space. The turn of motive is not important, if the result is good.

Island House,立面面向河流,避免了来自街道的目光
Island House. The facade fronts on the river, avoiding notice from the street.

至于我工作室的斜体书架有两个功能,其一是放书;其二是作为除了入口门之外能够引导人们往室内走的空间媒介,因此书架是斜体的。

TCK GoOnJai项目的阶梯花园和办公室公园项目类似的设计看似是为加强户外的三维效果。那些花园和KyeongDokJai项目的剧场性空间设计相关,那么这三栋房子的设计意图有什么相同或不同之处呢?

HMK GoOnJai办公室的阶梯式屋顶花园是风景式的小山。

办公园的看台式花园是一座建筑式小山,KyeongDokJai中音乐厅式的房间是一个室内的山形中庭,用来垂直连接各阶层的中心地带。

相同点是都是构造功能型地貌,不同之处在于GaOnJai项目是景观,办公园是建筑式小山,KyeongDokJai项目是空间式小山。

TCK 我觉得住户都非常满意,他们喜欢这房子甚至能理解其建筑意义。对业主而言,您的作品成果中什么是最重要的?您希望业主在每件作品中感受到什么?

HMK 建筑的持久性取决于使用建筑的满意度。

从设计到竣工全程,建筑师以他的专业性把控文化质量。但一旦入住,房屋的主人将以生活的痕迹来塑造自己的世界,这也将是房屋生命力的体现。

对业主来说,能感觉到我对每件作品的希望就是对设计理念的理解和认同。

通过对理念的认同,每一处空间都将被有效利用,随着时间的推移,建筑师的痕迹会褪去,而住户的生活痕迹的色彩将变得清晰。

因此,我认为一旦建筑内有人居住,该建筑才是真正的竣工,将会得到永生。

TCK 最后,请向我们大家说说您现在的想法、试验或兴趣。

HMK 我想研究多维空间,我本人对极端密集建筑的研究很感兴趣,因为这种建筑简单而丰富,又经济实用。我将继续在现代建筑中尝试韩式风格。

To sustain the shape of architecture, it should gain the necessity of existence. Once it losts the reason of existence, it should be deformed and be destroyed. Because function is life and shape is a form of life, the shape without function can't have life.

As for inclined book shelf of my studio, it has two roles, one is as book container, and the other is space media as an active guidance to inner space, beside the entrance door, so the book shelf is inclined.

TCK The stair-formed garden in GaOnJai and the similar one in Office Park seem to be designed to intensify the three-dimensional effect in the open outside area. Those gardens could be related to the theatrical space in KyeongDokJai. Then what is common and different in terms of the design intention in those three examples?

HMK Stepped roof garden in the office of GaOnJai is a small hill of landscape.

Stand garden of Office Park is a small hill of architecture and the concert hall-like space of KyeongDokJai is an interior hill-like atrium which is a central place of vertical linkage to various levels.

To construct functional topography is the same point and the different points are that GaOnJai is landscape, Office Park is architectural hill and KyeongDokJai is spatial hill.

TCK I got the impression that the residential clients were quite satisfied, enjoying the house and even understanding its architectural significance. What is the most important for the clients regarding the result of your work and what do you desire to be felt in each works?

HMK Sustainability of the architecture depends upon the satisfaction for using the architecture.

The architect manage the cultural quality by his professionalism from design stage to the end of construction stage, but after moving into the architecture, the owner cultivates his own world by stored trace of living which will be vitality of the architecture.

My hopes, to the owner, to be felt in each works of mine are understanding and sympathy on the design concept.

Through the sympathy on the concept, every space will be used efficiently, as time goes by, the color of architect will be faint, on the other hand, the colour of stored trace of living of inhabitant will be clear.

Therefore, once an architecture is completed by the living of inhabitant, the architecture will gain eternal life, I think.

TCK Lastly, please introduce your ongoing ideas, trials or interests to us.

HMK I would like to research on multi-dimensional spaces and I am interesting in study the extremely condensed architecture which is simple but rich, and economical.

I am trying to continue to experiment Koreaness in modern architecture.

GaOnJai项目,屋顶是曲形的
GaOnJai. The roof is curved.

KyeongDokJai项目,墙体是曲面的
KyeongDokJai. The wall is curved.

形式追随功能?功能追随形式!

由弗朗西斯·福特·科波拉导演的伟大战争影片《现代启示录》,讲述了主人公马丁·辛沿宽阔的湄公河而上,搜寻脱离美军的马龙·白兰度上校的故事。马丁·辛与几名士兵一路沿河而上,途中他们闯入了奇异的空间,目睹了形形色色的人和事。影片的线性叙事结构让观众可以十分轻松地紧跟剧情的发展,享受如画的风景。例如,影片中飞驰的船上,沿岸的风景在观众看来也是一种视觉享受,尽管情景让人害怕。然而,一旦你登陆,就会有一种不祥的预感,你的整个身体和周围环境都变成了可触可摸的体验空间。

建筑师金孝晚总是将"可视性"作为建筑设计的关键要素之一。这一概念表明他不认为建筑仅仅是地上的大楼,而把建筑结构看成一部电影,观众跟着它的流线性可以掌握整个故事。就像电影《现代启示录》的观众在故事讲述的整个过程都能知道自己身在何处,因为他知道故事的结点。金孝晚所构建的空间的使用者也总能感到整个空间是线性发展的空间,因此有不断拓展的空间感。而且,就像电影《现代启示录》中当你绕过河流看见的是新人和新景一样,当你沿设计师设计的路线四处走动时,你会进入新空间,面对不同的形式元素。沿线走走停停,住户感受到的是一个平面图像,而不是三维空间。然而当他偏离路线,进入空间,便出现立体感。扎哈·哈迪德也使用了这种线性结构,不同的是,她对路线上一切元素的设计都与整体设计和谐一致。而金孝晚则试图将路线与周围环境分离开来。正是这种分离,意外碰撞产生了。金孝晚的原创性就在于此。

功能追随形式

像Island House项目和WooNanJai项目一样,GaOnJai项目将房屋沿着边界排列以有效利用现场条件,从而使外围空间最大化。然

Form follows function? Function follows form!

One of the greatest war movies *Apocalypse Now*, directed by Francis Ford Coppola, unpacks its story of the protagonist Martin Sheen in pursuit of a deserted runaway officer Marlon Brando, going from the widely open downriver to the closely narrow upriver in a jungle and encountering the peculiar spaces and the various people and events with them. As the linear narrative structure allows the spectators to know when and where the ending comes right from the beginning, they can enjoy the development of story and the picturesqueness with easy relaxation. For example, the landscape around the river seen on the running boat can be visually appreciated like pictures despite of their uncanny scenes. However, once you make a landing, you feel ominous with your whole body and the surrounding environment turns into the space of bodily experiences.

The architect HyoMan Kim has always applied "the visibility" as one of the key elements in his architecture. This concept indicates that he denies the general concept of architecture which is a building stays on the land, but treats an architectural structure like a film whose whole story can be grasped when the viewer goes along the linear flow of it. As the spectator of *Apocalypse Now* becomes to know where he is in the whole process storytelling because he is aware of the ending point, the user of the space HyoMan Kim constructs always feels the whole space in the linearly developing space and thus uses it with the extended sense of space. Also, as you can see a new scene and meet new characters when you round the river like in the above movie, you can face another space and different formative element when you go around. Staying and seeing on the route, the user appreciates the space rather as a flat picture, not sensing it as a three-dimensional space. However, when he deviates from the route and goes into the space, then he experiences the three-dimensional feeling. Zaha Hadid

GaOnJai项目的办公室可由穿过室内庭院进入，人们在此每天都可感受到自然

entering through the innercourt, office in GaOnJai has the merit to feel the nature everyday.

而，比起Island House项目和WooNanJai项目，GaOnJai项目又前进了两步。即它的中央外部空间是分段分区的，这样它就被分为几个零碎空间，并以此作为基本材料。然后，建立一条从入口到最后一个房间的路，并将零碎空间安排在每一个拐点上，以此产生独特的场景。像上例影片一样，这一设计将场景沿途置于每处零碎空间内，使我们感受到空间的发挥，特别是一路上被命名的空间。事实上，这与大多数建筑师的方法并无大的差别。因为现实建筑作品的关键是解决差异与混乱的问题。换句话说，填补既定条件和理想结果之间的空白就是建筑思路的起点，试图把分散的线路联系起来则是形成构造的开始。

然而建筑师金孝晚在解决条件和结果的差异问题上以他独特的方式展现了他的睿智：他试图将既定路线和零碎空间碰撞在一起，凸显两者之间的空间差异，或者使它与设计路线框架背道而驰。为了同时解决这一问题，建筑师设计的原创形式产生了。尽管这些形式通常是对角线，以连接或填平空间间隙，事实上却产生了与路线设计不符且相当有创意的形式。位于西南至东北的这一对角线方向的客厅显示了这一设计原则。一开始方向是来解决功能性问题的，换句话说，反映朝向临近山顶的递减性。然而在路上一瞥，它表明建筑师旨在塑造一位肩膀上扬的骄傲的女演员形象，因为房屋呈对角线的形状位于公路附近。

此外，这幢房子采用了韩式风格的凹线形屋顶和屋檐。立面曲线与平面倾斜线之间的意外碰撞产生了外部的独特造型和内部的特有空间。

形式追随功能

KyeongDokJai项目的朝向则与GaOnJai项目相反，因为它相对

also uses this idea of linear route. Whereas she plans the elements on the route in a harmonious and consistent relationship with the overall design, Kim tries to separate the route from the surrounding spaces. From the separation, unexpected collisions break out. HyoMan Kim's originality lies in this.

Form follows function.

GaOnJai maximizes the outside space by laying out the house along the boundary in order to make effective use of the condition of the site as in the cases of Island House and WooNamJai. However, GaOnJai goes further to the next two steps unlike the past two houses mentioned above. That is, the central outside space guaranteed by the layout is segmented and separated so that it can be divided into several fragmented spaces which are prepared as the basic material. Then, a pathway route from the entrance to the final space is established and the prepared material is arranged in every curved point so as to make a unique scene in each point. Like in the exemplary movie, this design puts the formative scenes in each fragments of the pathway route toward the final point and allows us to feel the spatial play especially in the named places along the route. In fact, this is not that different from a majority of architects' method since the key in an actual architectural work is to resolve the problem of differences and dislocations. In other words, filling up the gap between the given conditions and the desired result is the starting point of architectural ideas and the intention of linking the dislocated lines is the beginning of constructing forms.

However, the architect Kim displays a witty wresting in his own unique ways of resolving the differences between the condition and the result: he tries to make the established route and the fragmented spaces collided to expose the spatial differences between the two or to make it go against the direction of the frame of route. In order to solve the problem simultaneously, there appear

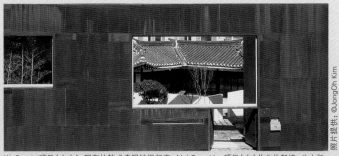

WaSunJai项目（左中），现存的韩式房屋被框起来，NokSungHun项目（右）作为构架墙，为人们提供观景的视野
WaSunJai(left, middle), the existing Korean house is framed, NokSungHun(right) as framed walls with the view of landscape

小一些，而且背南面北。GaOnJai项目采用的是内向式凹雕式布局，将房屋排列于场地边界处，以确保中央的外部空间，而KyeongDokJai项目，采用的是外向式布局，将房屋与外部空间置于三维立体空间中，以突出形式与空间的整体性。假设它的南部被邻房所围合，那么北面一定面向公共绿地，并且考虑到用户的采光需求，建筑师必须保证向北无论内外均有一定开阔的眼界。因此，房子呈扇形散开，为外向浮雕形式。

此外，为了克服场地面积的局限性，三维立体空间采用跃层结构，使之更富活力。就在能见度高的设计路线与三维空间的交汇点，层次感显露出来的地方，空间由主路分离而浮于四处。最好的例子就是颇具浮雕设计形式的室外亭子。相反，由于与主路相连，客厅变成了建筑师作品中最独特的剧场空间，而在这里最独特的要数作为覆层的铝屏，它是显示浮雕方向的关键元素。铝屏作为立面，面向北面公路，是建筑师作品中最具意义的尝试。虽然不一定需要它的覆盖功能，铝屏却积极扮演着门脸的形式功能。这一设计成为可能仅仅是因为整体设计依赖于浮雕的趋向。当然采用铝屏并不是他的第一次尝试，然而这个项目与现存案例在细节方面存在差异，那就是铝屏采用的是曲线而非对角线，并且曲线自身是三维的而不是二维的。由此，建筑师显露出他的后现代主义风格。

剧场空间的技巧

在我看来，KyeongDokJai项目有趣的元素是多层次的客厅，而不是吸引客户眼球的外观设计。由于空间限制，客厅成为带有许多视觉元素的分层空间。然而，更重要的是家庭成员在客厅里的多样活动。融洽与共融应该是设计住宅最为关心的要素。然而，房子应该是住户

the original forms the architect contrives. Although these forms, which are usually diagonal, are to connect or to smooth down the gaps, they result in quite provocative forms which are not adjusted to follow the frame of route. The direction of living room located diagonally from the southern-western road shows this design principle. The direction is decided to solve the functional condition at first, in other words, to reflect the axe toward the neighboring mountain top. Nonetheless, when giving a glance from the road, it exhibits the architect's pursuit of form looking like a proud actress with one of her shoulders rising because it is allocated diagonally to the road.

In addition, this house displays the trial of concave line of the roof and eaves as the Korean style. This design of an unexpected collision between the curved lines of elevation and the canted line of plan produces a unique formative outside and a peculiar space-scene inside.

Function follows form.

KyeongDokJai needs to be located in a different direction from GaOnJai since it is a comparatively small house and the area faces the north. Whereas GaOnJai shows the form of introvert intaglio arranging the house along the site borders so as to secure the outside space in the middle, KyeongDokJai, in contrast, shows the form of extravert emboss locating the house and the outside space in a three-dimensional way so as to highlight the form as well as the space as masses. Conditioned that the southern side is closed by the neighboring house and that the northern side is open to the public green area, and considering the client's demand of letting the sun in, the architect had to guarantee a visual openness from inside as well as outside with an open view to the north. As a result, the house becomes to fan out an extravert emboss form. And besides, in order to overcome the site's areal limitedness, the three-dimensional space is vitalized by applying skip floor. Thanks to where the planned route for visibility and the three-dimensional spaces meet and the difference of levels is exposed, spaces become to be separated from the route and float around. The most exemplary space is the outside pavilion standing out with its formativeness of emboss. On the contrary, the living room becomes a most unique theatrical space among the architect's works due to its combining with the route. Yet, the most unique here is the aluminum screen for covering, a key element of showing the direction of emboss. This screen as the facade, facing the road in the north side, seems to be a significant tryout among the architect's works since the screen is actively applied with its formative role as the face of house in spite that its covering function

WaSunJai项目，将老房子框起来
WaSunJai, framing the old house

NokSungHun项目，将景观框起来
NokSungHun, framing the landscape

可以娱乐和活动的地方。在KyeongDokJai项目中，这两个互相矛盾的目标可以同时实现。家庭联系并不总是需要共同参与活动来实现，同样也可以通过旁观他人活动来实现。父亲打鼓，母亲准备晚餐，儿子读书，女儿则听着鼓乐，仰望天空。这样的话，这处剧场空间就不是单向的，你可以从观众席看舞台，也可以从舞台看观众席。这种相互关系是剧场空间的必需条件。

办公园项目也有剧场式空间，在这儿主要是办公，但也不总是作为办公区。因此建筑师把它设计成俏皮的空间而非办公区。亭阁与假山矗立于中心，来作为带有一排坐椅的剧场空间。人们从这里可以看见附近的办公桌，和上述的例子一样，休息的工作人员可以在中心的剧场空间看见同事办公，相反，由于整个空间的开放性与可视性，办公人员也可以看见休息的同事。

模式和物体

由于办公园项目与Iroje KHM工作室都是办公区，因此功能性是设计首要考虑的。然而，建筑师通过运用雕像和地面实现了他对形式表现的愿望：首先，他通过框架打造了一块平地，这不仅实现了它的基本功能，而且人们可以在上面作画。其次，他将物体作为表现元素与平地形成鲜明对比。GaOnJai等项目的概念让我们通过眼睛与其他感官来传达充满活力的、不断变化的电影式场景，在这里，我们沿着规划的路线，来偶遇一些特殊的场景，而办公园则允许我们以静止状态瞥见整个计划。这一设计似乎有两种效果：形式效果，即整体景象立马呈现，因为形式和物体并不互相遮掩。然而，两种不同形式元素之间的碰撞被延展为两种不同活动之间的视觉碰撞。因此，这种碰撞产生一种模糊性，它表明建筑师想要将娱乐与工作合二为一的意图。这

is not necessarily needed. This design is possible only because the overall design is dependent on the tendency of emboss. Surely this element is not the first tryout of his, but it is different in details from the existent example in that the curves are applied instead of the diagonal lines and that the curves themselves are not two-dimensional but three-dimensional. With this, the architect began to reveal his postmodernist touch.

Playfulness of the theatrical space

To my eyes, the interesting element in KyeongDokJai is the multi-leveled living room rather than the formative exterior, where the client's discerning eyes played a big part. Due to the areal limitedness of the site, the living room becomes the place of collecting all the layered spaces with many visual elements. However, it is more important that the activities of the family members are layered in the living room. The rapport and communion should be concerned most in designing a residential house. However, a house should be a place where the residents could have their own enjoyments and activities. In KyeongDokJai, these two contradictory goals can be achieved simultaneously. The family bonds can be shared not always by participating together in something but by having the others' activities around and looking at each other. One can play drum and the other prepares dinner while the son reads a book and the daughter looks up the sky listening to the drum playing. That way, the theatrical space cannot be one way. You can have a look at the stage from the audience seats and you can also give a look at the audience. This mutuality is the necessary condition of the theatrical space.

The Office Park has also a theatrical space. This place is mainly for business but is not always used as an office, so the architect makes it as a playful space rather than a workplace. The pavilion and the so-called hill, standing like an object in the center, work as a theatrical space with tiers of seats. From this space, the work desks nearby can be seen. Like in the above example, the staffs can take a rest in the central theatrical space watching the working colleagues and in reverse, the working staffs can see the relaxing colleagues thanks to the openness and visibility of the whole space.

Patterns and Objects

Since the Office Park and Iroje KHM Studio are workplaces, the primary concern of design should be functional. Nonetheless, the architect realizes his desire of expressing the formativeness by the method of figure and ground: firstly, he creates a ground by applying the framing patterns, which not only fulfills the basic function but also works as a ground on which a formative picture can be drawn; then secondly, he puts an object as an expressive ele-

办公室公园1和2,室内应用了可调整的格局以及各种造型的物件
Office Park 1 and 2, adjusted patterns and formative objects were used for the office interior

样,金孝晚展现了他作为一个后现代建筑师的内在特点。此外,Iroje KHM工作室自由且真实地表现了其模式的形式,这一形式并没有代表某些功能,反而是向天花板倾斜和延展,如同飞翔一样,没有被其原始功能所干扰。倾斜的书架在功能上是没有问题的,其丰富的形式也足够有效。

给建筑师的话

韩国人对流行元素非常敏感,相应地韩国文化变化很快。然而这些变化似乎不是对某一不同特征的审判,而是对某一起源于某地的流行趋势的强制适应。我认为我们热衷于流行是因为痴迷于随大流。过去,当工业化等同于经济发展的时候,大规模生产意味着令人艳羡的发展,即使只有很少的数量,但只要你可以利用相似的产品与他人联合,你就可以伴装主流。相比之下,独特性因为非主流而被视为古怪恶劣。因此人们想要与他人相似,且隶属于同一集体。这就是为什么公寓楼成为居住的主要形式的原因。因此主流建筑风格取决于现代化大生产。现代建筑的基本原则,用勒·柯布西耶的话说就是"用来盛装生活的机器"。逻辑功能可以解释一切结果,相应地结果也应是逻辑推导的那样。毕竟,建筑的功能性降到了追求箱形而排斥装饰性。然而,为了填补由于没有装饰而造成的审美缺陷,又不得不求助于几何逻辑、抽象理念与精英美学。在相同情况下,去除装饰要求建筑师给予解释和逻辑性,而不是形式感观。因此,蹩脚的形式设计很容易让人瞧不上。然而现在,越来越多的人,只要买得起,都希望能够拥有田园式而不是公寓式的住房。看起来,人们享受的是非主流而不是主流的生活方式,想买的是手工制作而非大批量生产的日常用品。手工艺品粗糙却富于感性和创造性,因为手工艺者在制作过程中关注的是他

ment to make a pictorial contrast on the ground. Whereas the concept of GaOnJai and etc. delivers the dynamically changing filmic scenes moment by moment felt through your eyes and other senses where you move along the planned route and encounter some special scenes on the way, the concept of the Office Park allows us to feel the whole plan like a static picture at a glance. This design seems to have two effects: the effect of formativeness is that the entire view is available at once because the pattern and the object don't mask each other; but still, the formative collision between the two different formative elements can be extended to the visual collision between the two different activities. As a result, the collision causes a kind of ambiguity, which reveals the architect's intention of hybridity of scenes including both playful activities and working activities. With this, HyoMan Kim exposes his inner characteristics as a postmodern architect. Besides, Iroje KHM Studio is free and frank to express the formativeness of the patterns, not representing some functions but being canted or stretched toward the ceiling like flying without being interfered by their original functions. The slanted bookshelves have no problem of function, but their expressiveness formativeness is effective enough.

What is remained to tell the architect

Koreans are very sensitive to the popularity and accordingly Korean culture changes rapidly. However, this change seems not a new trial to claim different characteristics for each but a forced adjustment to the ongoing popularity arrived from somewhere. I think we are keen to the popularity because we are obsessed to belong to the majority. In the old days when industrialization was identified with the economic development, mass production meant development which is desirable. Even with small amount you could pretend to be major only if you can be united and padded out with the others with similar things. In contrast, the uniqueness was treated as weird and bad as it was not major. Therefore, people wanted to resemble each other and belong to the same group. That might be why apartment houses became the main form of residence. Thus the leading architectural style has been dependent on the modernism of mass production. The basic principle of modernist architecture is to make "a machine to contain life" in Le Corbusier's terms. Every result is explained by the logic of function and accordingly the result is asserted to be necessarily as it should be. After all, the functionality in architecture is reduced to the pursuit of box-shape and the exclusion of decoration. However, the geometrical logics are added and the abstract and elitist aesthetics are forced in order to fill up the aesthetic deficiency caused by the absence of decoration. In the

Iroje KHM 建筑师事务所工作室的书架墙
Iroje KHM Architects Studio bookshelf wall

们的感觉。情感源自于身体本身而非逻辑，因此所产生的结果的任何一方面都是无法解释的，也是不应该解释的。

　　建筑师金孝晚师从于SooGuen Kim（20世纪70至80年代的主要建筑师），所以他深受现代主义的影响，致力于从初始便寻求功能性的箱形结构。他的设计方法的发展总是以功能逻辑和建筑理念为基础，且其形式主要取决于实际的功能。然而，在设计过程中，他并没有受到初始的设计原则的限制，相反，他还提出了具有争议性的设计方式，并且通过细部构件间的差错，来形成差异。这些差异通过具有象征性的冲突来产生独特的结果：它们展示了一些装饰性元素，这些元素通过对角线突出了室外的造型和室内舞台独特的场景。同样的，利用矛盾的方法来解决差异性也是这位建筑师趋向于采用的方法。也就是说，他并不是以现代主义的建造过程为基础，来积极地表达建筑形式，而是通过他自己设计的逻辑，使其形式被动地形成。这一方法旨在展示建筑师本身的风格，我将此定义为"手工艺"。这从建筑师的草图秀以及其他介绍其处理形式的方法的笔记中都可以看出。

　　总之，金孝晚是矛盾的：他的建筑本质赋予他自由，以追求形式和感性，但是作为一名现代主义建筑师，他仍然试图在理性的基础上来表达感性。

　　让我们将其总结为现代主义装束内的一颗后现代主义心灵，手工艺品并不是过时的，现在正在成为一个重要的里程碑。不是每件事物都要得到解读，比如说过去年代里那些著名的画作，你真的希望在这些画作中看到了一些逻辑性的阐释吗？为了享受到更多自然赋予的自由，我认为是时候去扭转路易斯·沙利文的观念了。形式追随功能？功能追随形式！

same context, the elimination of decoration asks architects to give explanations and logics instead of the formative senses. Thus, the unexplainable formativeness is easily treated as inferior. Nowadays, however, more and more people hope to have their own form of residences like idyllic houses instead of apartment houses only if they can afford. Seemingly, people want to enjoy the minor lifestyles rather than the major ones and desire craft goods rather than mass-produced commodities. Craft is rough but sensible and creative because the crafty artists focus on their senses while working. Sensibility is brought out not from the logics but from the body itself. Therefore, every aspect of the result is not explainable and should not be explainable.

Architect HyoMan Kim has learned from SooGuen Kim, the major architect in the 1970~1980s, so he was under the modernist influences seeking the functional box-shape from the outset. His method of design development is always based on the logic of function and his architectural ideas and formativeness depend fundamentally on the realistic functionality. Nevertheless, he is not subjugated by the starting principles in the proceeding process. On the contrary, he advances to the provocative way and creates differences between the detailed elements by making them going amiss. These differences lead to the unique result through the indicative clashes: they show the decorative elements emphasizing the exterior formativeness and stage unique scenes inside by the diagonal lines. As such, the way of resolving the differences with contradictory methodology is intended by the architect. That is to say, he does not actively express the form based on the modernist process, but let the form be passively achieved by the planned logic of his. The goal of this methodology is to reveal the architectural color of his, which I would like to define as "the handcrafted" What his sketches show and what the other journals introduce in terms of his method of dealing with the formativeness prove that. In conclusion, HyoMan Kim is ambivalent: his architectural nature gives him the freedom to pursue the formativeness and sensibility, but still he makes attempts to express the sensibility on the basis of rationality as one of the modernist generation architects. Let's say: a postmodernist heart in modernist outfits. The handcrafted is not out of date but is becoming an important milestone now. Everything should not always be explained. Let's watch the famous paintings in the past. Do you really need a logical elucidation in those artworks? In order to enjoy more freedom by nature, I think it's time to reverse Louis Sullivan's concepts. Form follows function? Function Follows Form! *TaeCheol Kim*

金孝晚 — KyoMan Kim
GaOnJai

纪念性场地

大约在20年前，政府在GangNam举办了一场"住宅展览会"，并且建造了一个展览住宅小镇，其内的建筑均为低层住宅。小镇的每处场地都是由选定的建筑师来进行设计，他们在那个年代的韩国都是非常著名的。

之前存在的本项目位于这座纪念性的小镇中。业主要求拆除掉现有的建筑，同时在原有场地内建造一座新的建筑。

甲方的要求

——其中最重要的一个要求是尽可能地提高土地利用率，并且使生态空间最大化。这是因为业主认为拆除的建筑的室外空间太过于狭小，导致室内空间相当的黑暗。

——住宅的基本空间即是"内向型空间"，这种空间具有很多优点：保持私密性，提高安全性能，使其免受周围环境的侵扰。

——起居室拥有充满活力的、风景如画的视野，这些视野可以媲美附近山体的优美景观。

文化认同

这里要求住宅本身带有自己的韩国文化属性，因为这个小镇是韩国最理想的住宅小镇之一。

将传统的语言"MaDang""Ru""CheoMa""DolDam"传承下去

通过引入"MaDang"一词，即韩式室内庭院，建筑师能够建造"内向型"空间、生态环境与有用的室外空间。通过引入"Ru"，即韩式底层架空式建筑，建筑师能将附近山体动态和风景如画的景观展现出来，并且把起居空间的轴线转变成类似于周围山体的顶部。"Ru"建筑的上空空间成为这座建筑内具有戏剧性的一系列漫步路程中的起点，如同洞穴一样。

此外，通过引进曲面的"CheoMa"，即悬挑的韩式屋顶，来作为这座住宅中浮动的屋顶，建筑师不仅仅能够抵抗恶劣的天气条件，同时还能复兴传统的韩国建筑语言。

通过引进椭圆形的混凝土墙体（起源于传统的石墙），建筑师希望它成为历史的见证。

项目的地形布局

通过适应现有场地的地形，位于较低和较高位置的MaDang以及各种跃层式的室内空间应用在了项目的不同楼层中。它们在住宅的每个角落都产生了有趣的漫步路线。

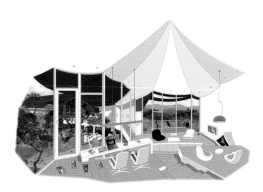

这种对于自然来说所具有的天然适应性曾是韩国建筑文化中的基本特点。

漫步

住宅中的每处空间都在流动着，来作为参观路线。这条路线通过室内和室外空间，在视觉上和空间上都强烈地引起了人们漫步的兴趣。

因此，建筑师希望住宅内家庭成员在其间生活时，尤其是他们在自己的世界内漫步时，再也不会感到枯燥。

项目名称：GaOnJai
地点：BunDang-gu, SungNam-si, GyeongGi-do, Korea
建筑师：HyoMan Kim
项目团队：KyungJin Jung, SeungHee Song, SuKyung Jang, JiYeon Kim, EunJin Shin, HyeJin Kim, WooSin Sim
承包商：Jehyo 用途：residence
用地面积：643.5m² 总建筑面积：319.96m² 有效楼层面积：329.35m²
结构：concrete rahmen
室外饰面：black zinc plate, white stucco, exposed concrete
室内饰面：lacquer, wood flooring, perforated steel plate
设计时间：2010.2—6 施工时间：2010.9 竣工时间：2012.5
摄影师：©JongOh Kim

151

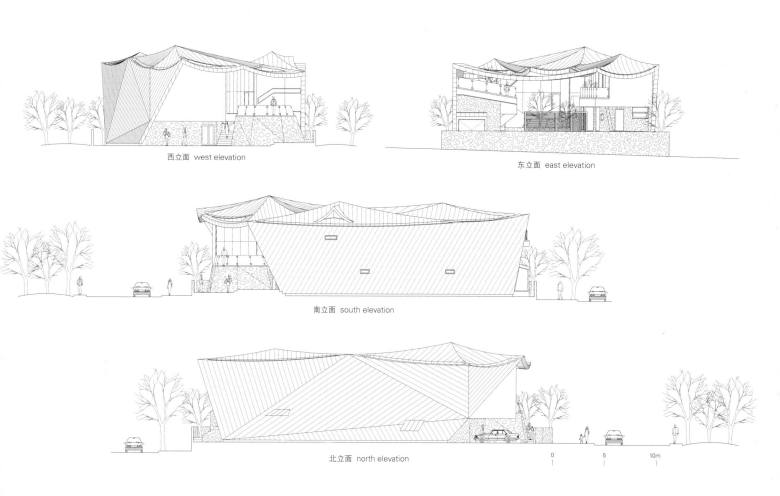

西立面 west elevation

东立面 east elevation

南立面 south elevation

北立面 north elevation

0　5　10m

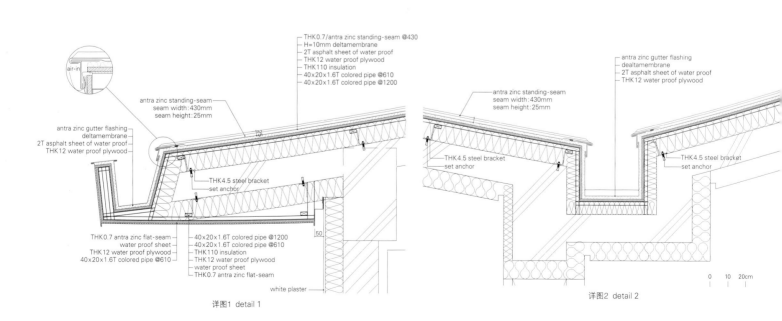

详图1 detail 1　　　　　　　　　详图2 detail 2

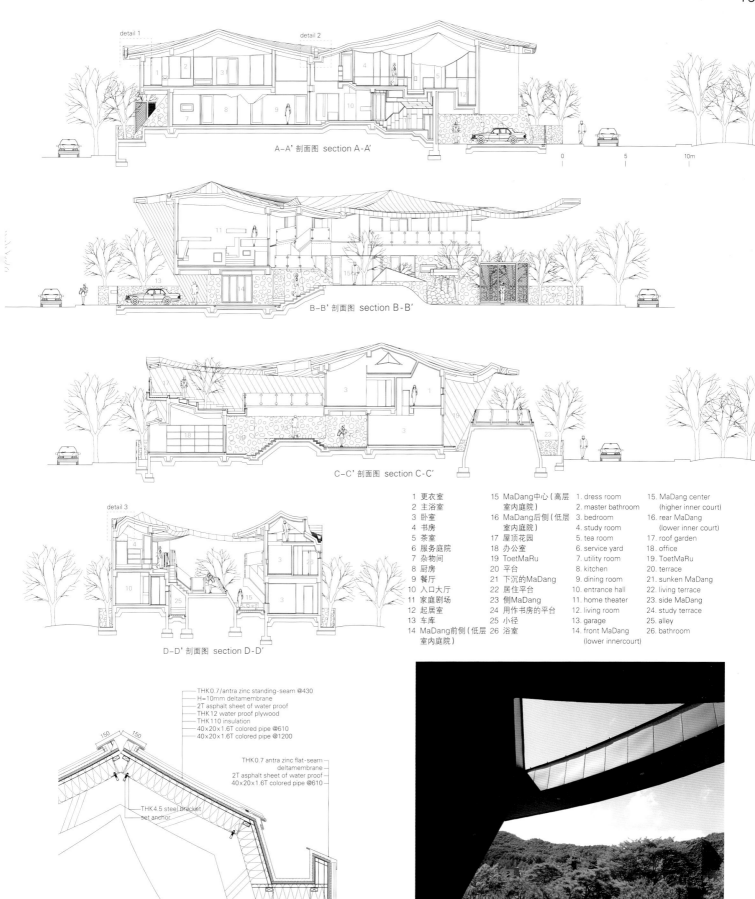

A-A' 剖面图 section A-A'

B-B' 剖面图 section B-B'

C-C' 剖面图 section C-C'

D-D' 剖面图 section D-D'

1 更衣室	15 MaDang中心（高层室内庭院）	1. dress room	15. MaDang center (higher inner court)
2 主浴室	16 MaDang后侧（低层室内庭院）	2. master bathroom	16. rear MaDang (lower inner court)
3 卧室		3. bedroom	
4 书房	17 屋顶花园	4. study room	17. roof garden
5 茶室	18 办公室	5. tea room	18. office
6 服务庭院	19 ToetMaRu	6. service yard	19. ToetMaRu
7 杂物间	20 平台	7. utility room	20. terrace
8 厨房	21 下沉的MaDang	8. kitchen	21. sunken MaDang
9 餐厅	22 居住平台	9. dining room	22. living terrace
10 入口大厅	23 侧MaDang	10. entrance hall	23. side MaDang
11 家庭剧场	24 用作书房的平台	11. home theater	24. study terrace
12 起居室	25 小径	12. living room	25. alley
13 车库	26 浴室	13. garage	26. bathroom
14 MaDang前侧（低层室内庭院）		14. front MaDang (lower innercourt)	

详图3 detail 3

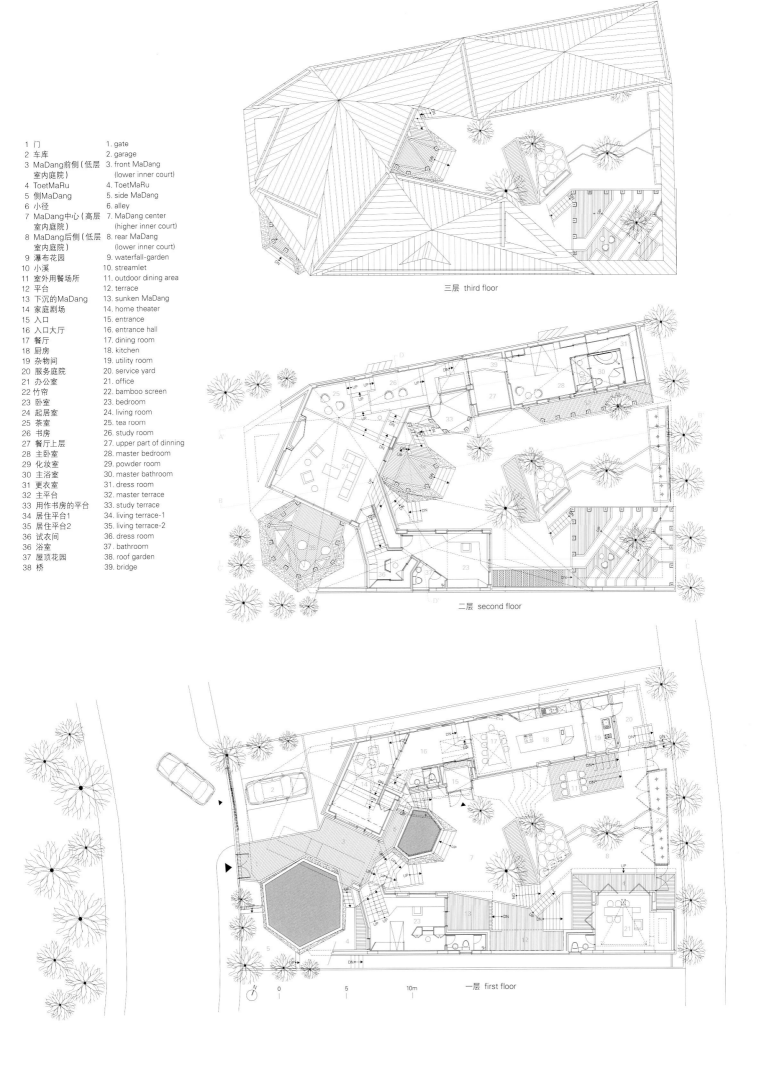

1 门	1. gate
2 车库	2. garage
3 MaDang前侧（低层室内庭院）	3. front MaDang (lower inner court)
4 ToetMaRu	4. ToetMaRu
5 侧MaDang	5. side MaDang
6 小径	6. alley
7 MaDang中心（高层室内庭院）	7. MaDang center (higher inner court)
8 MaDang后侧（低层室内庭院）	8. rear MaDang (lower inner court)
9 瀑布花园	9. waterfall-garden
10 小溪	10. streamlet
11 室外用餐场所	11. outdoor dining area
12 平台	12. terrace
13 下沉的MaDang	13. sunken MaDang
14 家庭剧场	14. home theater
15 入口	15. entrance
16 入口大厅	16. entrance hall
17 餐厅	17. dining room
18 厨房	18. kitchen
19 杂物间	19. utility room
20 服务庭院	20. service yard
21 办公室	21. office
22 竹帘	22. bamboo screen
23 卧室	23. bedroom
24 起居室	24. living room
25 茶室	25. tea room
26 书房	26. study room
27 餐厅上层	27. upper part of dinning
28 主卧室	28. master bedroom
29 化妆室	29. powder room
30 主浴室	30. master bathroom
31 更衣室	31. dress room
32 主平台	32. master terrace
33 用作书房的平台	33. study terrace
34 居住平台1	34. living terrace-1
35 居住平台2	35. living terrace-2
36 试衣间	36. dress room
36 浴室	37. bathroom
37 屋顶花园	38. roof garden
38 桥	39. bridge

三层 third floor

二层 second floor

一层 first floor

GaOnJai
Monumental site

About 20 years ago, the government held "the house expo" in GangNam and constructed this "expo town" of low-rise residences. Every site of this town was designed by selected architects who were famous at that time in Korea.
The former building was also in this monumental town. What the owner requested was to demolish the existing building and construct a new one at the same site.

Requirements from the client

– One of the important requirements was maximizing the efficiency of land use and ecological spaces. This was because the owner thought that the exterior space of demolished old house was too small and the circumstance of inner space was relatively dark.
– The essential space for this house was the "introvert space", as this kind of space has a lot of advantages: keeping privacy, improving security, and protecting from surrounding environment.
– The living room with dynamic and picturesque view which resambles the landscape of nearby mountain was also required.

Cultural identity

The house with its own identity of Korean culture was required as this town is one of the most desirable residential towns in Korea.

Passing down traditional language "MaDang", "Ru", "CheoMa", "DolDam".

By introducing "MaDang", Korean inner court, the architects could create introvert space, ecological environment, and useful outdoor space.
By introducing "Ru", Korean-pilotied architecture, the architects could reflect the dynamic and picturesque landscape of nearby mountain, turning the axis of living room to top of the surrounding mountain alike. Void space under "Ru" is the beginning part of dramatic sequence of strolling course in this house, just like cave. Besides, by introducing curved "CheoMa", cantilevered roof of Korea for the floating roof of this house, the architects could obtain not only the resistance to rough weather condition, but also the revival of Korean traditional architecture language.
By introducing oval patterned concrete wall, derived from traditional stone wall, the architects expected that it would be a reminder of the past.

Topographical layout of programs

By adapting topography of existing site, lower MaDang, higher MaDang, and various skip-floored inner spaces were designed in different levels. They produced interesting routes of strolling at every place of this house.
This natural adaptation to nature has been a basic character of Korean architectural culture.

Strolling

Every space in this house flows as the touring course that has dramatic interest of "strolling" through the inside and outside of this house, both visually and spatially.
Thereby the architects expect that the family members will never get bored while living in this house in particular as they stroll in their own world. HyoMan Kim

KyeongDokJai

薄片式扇形布局——在冬天从南部引入阳光

建筑师选择了一个生态友好型布局,即由三个薄片式扇形区域构成。它包含一个南侧轴线,来把冬季的阳光引入住宅室内和室外空间。

小型场地内的花园空间最大化

在这片面积为230m²的场地内,建筑师试图将花园的面积最大化,并且使花园具有多样性,包括试点花园、室内花园、试点亭阁市花园、下沉花园以及屋顶花园。

环保型围栏立面——过滤夏天的阳光和噪音,保持私密性和安全性

建筑师将建筑的前立面采用铝管围栏来覆盖,以过滤炎热的夏季光线和噪音,保持私密性和安全性。

传统景观——浮动与漫步

浮动的景观——白色铝管构成的围栏具有动感的曲线,成为建筑的室外表皮。这一设计来源于传统的屋顶"CheoMa"。多孔混凝土围栏成为传统石质围栏的隐喻。

房间内的每一处空间都如同一次旅程,人们在室内外穿梭的时候,在视觉性和空间性方面,都能在漫步时产生极大的愉悦感。

家庭规划——音乐厅式的空间

为了增进家庭成员之间的感情,建筑师设计了"一处复杂的空间",它如同一个音乐厅,在二层设置起居室以及餐厅,而在楼梯层和三层则设置书房、儿童卧室、家庭式酒吧以及置放了钢琴和鼓的舞台。因此,所有的家庭成员都能在日常生活中经常互动。

非建筑景观

白色铝管围栏发挥着重要的作用,即制造住宅内的各种拼贴构件,它们是透明的,且非建筑材质的。这一住宅体块与其他类似的房屋并排设置,因此,非建筑性的景观被添加进来。

项目名称:KyeongDokJai
地点:1108-4, HaengSin-dong, DeogYang-gu, GoYang-si, GyeongGi-do, Korea
建筑师:HyoMan Kim
项目团队:KyungJin Jung, SeungHee Song, SuKyung Jang, EunJin Sin, HyeJin Kim
用途:residence 用地面积:236.30m²
总建筑面积:115.68m² 有效楼层面积:329.66m²
结构:concrete rahmen
室外饰面:dryvit, aluminum pipe, glass
室内饰面:confloor, wood flooring, V.P, urethane paint on steel plate
设计时间:2010.3—7 施工时间:2010.9 竣工时间:2012.2
摄影师:©JongOh Kim

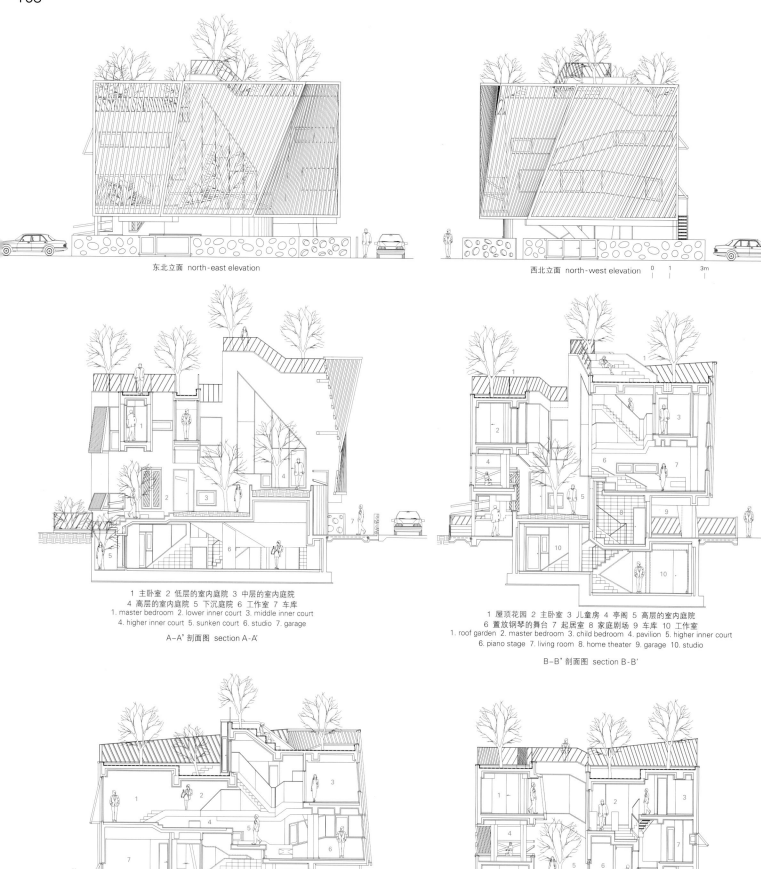

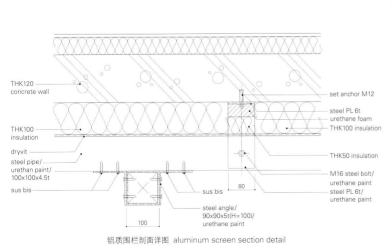

铝质围栏剖面详图 aluminum screen section detail

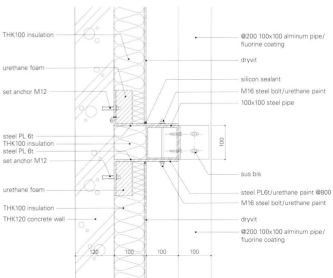

铝质围栏水平剖面详图 aluminum screen hotizontal section detail

KyeongDokJai

Sliced pie-shaped layout – for introducing sunlight from the south in winter

The architects selected an eco-friendly layout which was composed by three sliced pie-shaped zones. It contains a southern axis to introduce winter sunlight into both inner and exterior space of the house.

Maximization of the garden space in the small site

In this small site, about 230m², the architects tried to maximize the area of garden and characterize the variety of gardens such as piloted approach garden, inner garden, piloted pavilion garden, sunken garden, and roof garden.

Eco-screen Facade – to filter summer sunlight and noise, and to keep privacy and security

The architects covered the front facade with aluminum pipe screen to filter the hot summer sunlight and noise for privacy and crime prevention.

Traditon scaping – floating & strolling

Floating shape – Dynamic curve of white aluminum pipe-screen as exterior skin is derived from the traditional roof, "CheoMa". And the perforated concrete fence is a metaphor of the traditional stone fence.

Every space in this house flows like a touring course, giving the dramatic pleasure of "strolling" through the inside and outside of this house, both visually and spatially.

Program for family – concert hall-like space

In order to add up a sense of fellowship between family members, The architects designed "one complex space" like a concert hall containing living room and dining room on the second floor, and study room, children's bedrooms, bar, piano and drum stage on the stair floor and third floor. Thereby, all members of the family always interact each other in daily life.

Non-architectural landscape

The white-curved aluminum pipe screen plays a role of producing collage of various elements of this house, translucent and non-architectural. This residential block is lined with similar houses, consequently, non-architectural landscape comes to be added .
HyoMan Kim

办公园

工作、休息与散步意外共存

随着整天呆在办公室里的员工的数量逐渐减少，现有的办公空间的效率也在降低。

为了解决这一问题，管理部门负责人决定通过提供一处有趣的、用于休息和漫步的混合空间，来最大化地利用空置面积。

由亭子和小山构成的景观

建筑师参考了"亭子"与"ToetMaRu"以及折叠的悬浮门来设计会议室。沿着"亭子"和用作图书馆的小山而设，这间办公室便不再是一座建筑，从隐喻的角度来说，则是一座韩式景观花园。

通过将公园与工作场所的概念相结合，这一项目绘制出了阶梯式花园、亭子、酒吧、休息室等功能。每个地方都被各种类型的工作和活

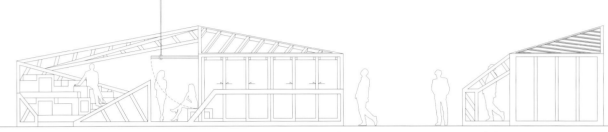

立面1_办公园1 elevation 1_office park 1   立面2_办公园1 elevation 2_office park 1

动赋予特色。

建筑师希望新型办公景观通过工作之间的互动及工作、休息和漫步的职能转换,将有助于提高效率。

传统与现代化之间具有对比性的和谐

年轻人,即该空间的使用者,要求将木材作为主材料,使其具有环保特性。通过应用透明的聚碳酸酯玻璃来作为隔间材料,办公室景观趋向于传统,但是通过木材与塑料玻璃之间具有对比性的和谐共处,该项目又极为现代化。

为了达到低成本建造,且突出木材性能的这一目标,建筑师在现有的墙体和天花板上覆上一层水泥砂浆,并且希望水泥砂浆较为庄重的现代性能够突出木材具有的经典的温和性。

立面3_办公园1 elevation 3_office park 1

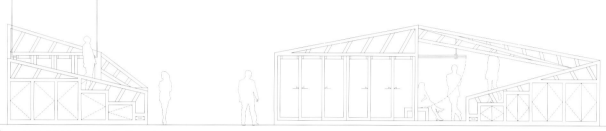
立面4_办公园1 elevation 4_office park 1

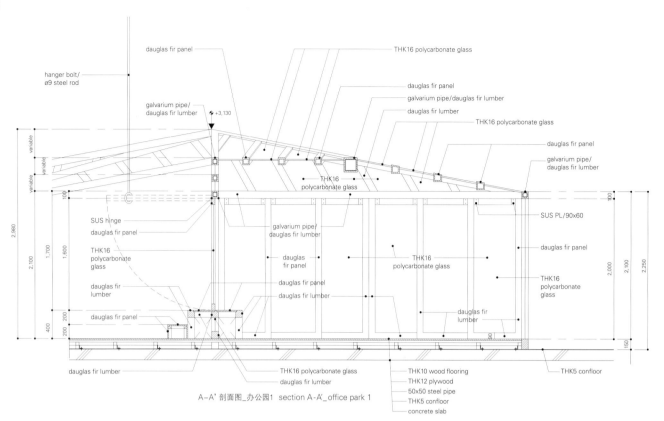

A-A' 剖面图_办公园1 section A-A'_office park 1

项目名称：Office Park
地点：Yeouido-dong, YeongDeungPo-gu, Seoul, Korea
建筑师：HyoMan Kim
项目团队：KyungJin Jung, SeungHee Song, SuKyung Jang
总承包商：Jehyo
用途：office
楼层面积：office park-1_304.46m² / office park-2_240.16m²
室内饰面：floor_confloor, vinyl carpet/
wall_confloor, V.P, dauglas fir plywood panel /
ceiling_confloor, V.P, expanded metal panel/
furniture_dauglas fir plywood panel, polycarbonate glass, tempered glass of frost, sliced veneer on MDF panel, fabric
设计时间：2011.6 竣工时间：2011.8
摄影师：©JongOh Kim

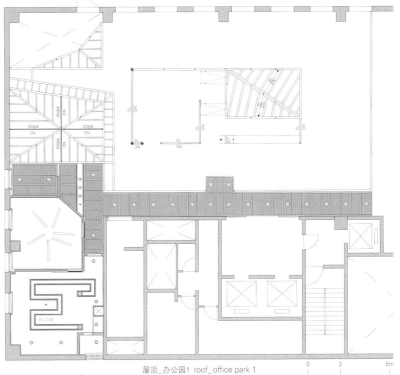

屋顶_办公园1 roof_office park 1

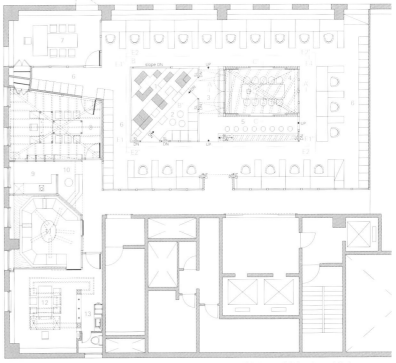

1 阶梯公园 2 休息室 3 街道 4 亭阁式会议室 5 酒吧 6 储物间 7 总监办公室 8 主任室2
9 厨房 10 信息台 11 接待室 12 主任室1 13 化妆间
1. stepped park 2. lounge 3. street 4. JeongJa conference room 5. bar 6. locker 7. chief room 8. director room-2
9. kitchen 10. information desk 11. reception room 12. director room-1 13. powder room
一层_办公园1 first floor_office park 1

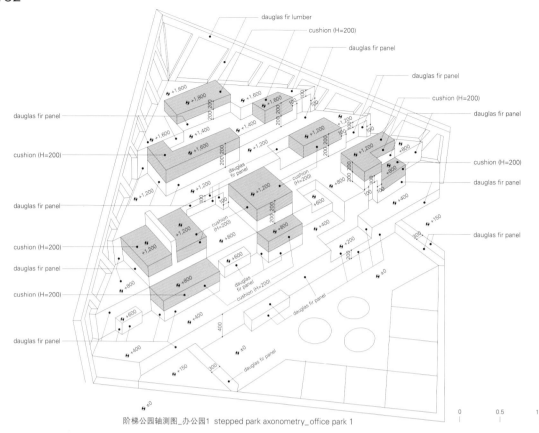

阶梯公园轴测图_办公园1 stepped park axonometry_ office park 1

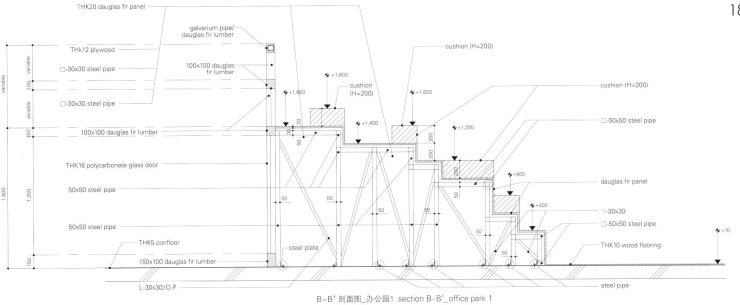

B-B' 剖面图_办公园1 section B-B'_office park 1

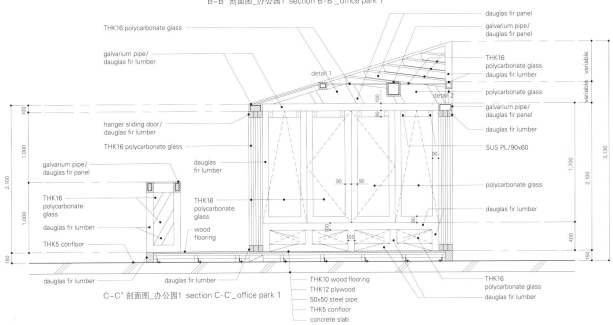

C-C' 剖面图_办公园1 section C-C'_office park 1

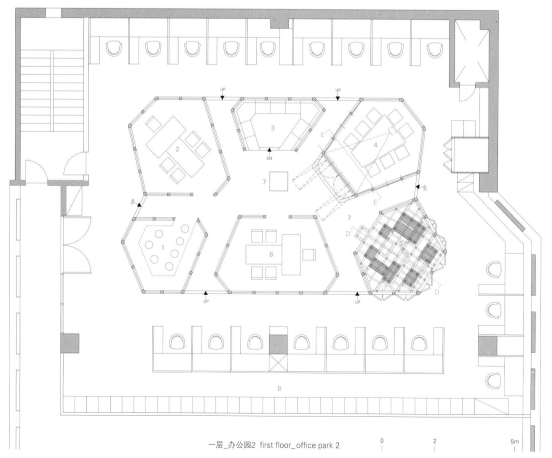

一层_办公园2 first floor_office park 2

1 酒吧
2 总监办公室1
3 休息室
4 亭阁式会议室
5 图书馆/阶梯公园
6 总监办公室2
7 广场
8 储物间

1. bar
2. chief room-1
3. lounge
4. JeongJa conference room
5. library/stepped park
6. chief room-2
7. plaza
8. locker

Office Park

Unexpectedness of coexistent of working, resting, strolling

As the number of employees staying all day in the office gradually decreases, existing work space gets less efficienct.
In order to solve this problem, the president of management decided to make the best of the vacated area by offering an interesting hybrid place for resting and strolling.

Landscape composed by JeongJa and small hill

The architects designed the conference room with reference to "JeongJa" with "ToetMaRu" and suspended-folding doors. Along with "JeongJa" and the hill for Library use, the office is no longer a building but Korean garden landscape, metaphorically.

By blending the concept of park and that of work place, the programs of stepped garden, JeongJa, bar, lounge, etc., were mapped out. Each place was characterized by the type of each kind of work or behavior.

The architects expect the new office landscape will help improve the efficiency with the interaction between characterized works and with the changeover between working, resting and strolling.

Contradistinctive harmony of tradition and modernity

Young people, the user of this space, requested to use wood for main material for its ecological properties. By adding translucent polycarbonate glass as a sort of partition, the office landscape turned out to be traditional but modern by contradistinctive harmony of wood and plastic glass.

To achieve low cost construction and to emphasize the property of wood, the architects covered the existing wall and ceiling with cement mortar and expected that stately modernity of cement mortar emphasizes the classic mildness of wood. HyoMan Kim

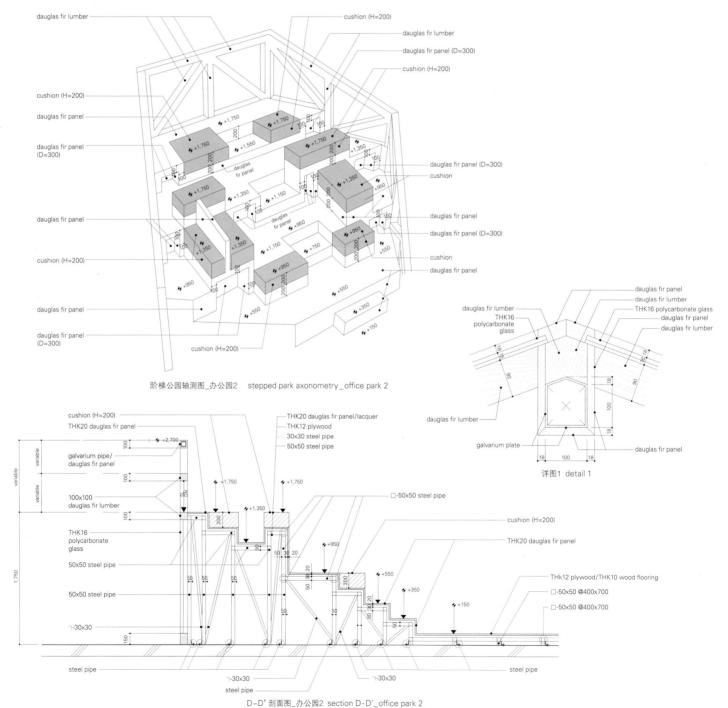

阶梯公园轴测图_办公园2 stepped park axonometry_office park 2

详图1 detail 1

D-D'剖面图_办公园2 section D-D'_office park 2

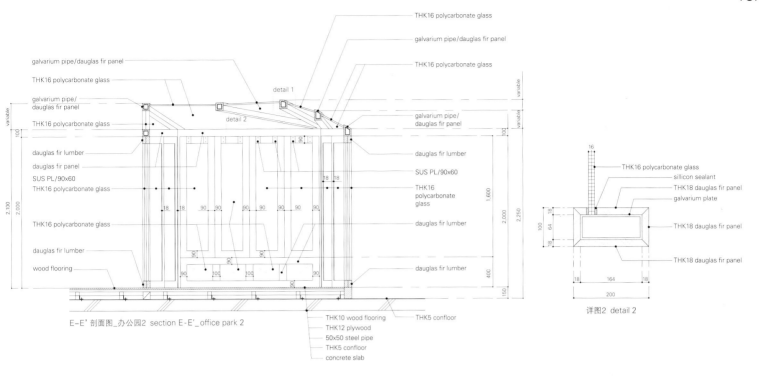

>>68
Archi-Union Architects
Was founded in 2003 by Philip Feng Yuan. Is a Shanghai based architectural design firm which is a grade A design qualification certified by Ministry of Housing and Urban-Rural Development(MHURD). Has managed to introduce a design style that is an amalgam of the global trends and the local traditional architectural approach. Philip Feng Yuan is an associate professor from Architecture and Urban Planning Institute of Tonfji University in Shanghai and co-founder of Digital Architectural Design Association(DADA) in China. He has received several architectural awards such as WA Chinese Architecture Award(2013), UED Best Museum Design Award(2013). Also he used to work as guest editor and has published many articles on several architectural magazines.

>>50
Nuntapong Yindeekhun + Bunphot Wasukree
Nuntapong Yindeekhun was born in Bangkok, Thailand in 1973. Received a Architectural Bachelor's Degree from King Mongkut's Institute of Technology Ladkrabang(KMITL), Bangkok in 1997. Bunphot Wasukree was born in Trat, Thailand in 1975. Received a Architectural Bachelor's Degree from KMITL, Bangkok in 1997.

>>88
a21 Studio
Is a small group of designers established in 2009. Has been leaded by Hiep Hoa Nguyen since then. They wish to bring their conception of life to the surroundings by architecture.

>>74
TYIN tegnestue Architects
Is a non-profit organization working humanitarian through architecture. Was established in 2008 and has completed several projects in developing areas of Thailand, Burma, Haiti and Uganda. Is currently ran by Andreas Grotvedt Gjertsen[right] and Yashar Hanstad[left] and has its headquarters in the Norwegian city of Trondheim. Has won several international awards such as Global Award for Sustainable Architecture(2012), WA Awards 10th Cycle(2012). Their projects have been published and exhibited worldwide. They aim to build strategic projects that can improve the lives of who are in difficult situations. The projects, realized through extensive collaboration with local residents and mutual learning, they hope, will mean beyond the physical structures.

© Pasi Aalto

>>40
Vo Trong Nghia + Takashi Niwa
Vo Trong Nghia was born in Vietnam in 1976. Graduated from Nagoya Institute of Technology in 2002. Received a master's degree from Tokyo University in 2004. Established Vo Trong Nghia Architects in 2006.
Takashi Niwa was born in Japan in 1979. Received a Bachelor of Architecture from Tokyo Metropolitan University in 2001 and Master of Engineering in 2003. Worked in Noriaki Okabe Architecture Network from 2005 to 2010. Has been a partner of Vo Trong Nghia Architects since 2010.

>>108
Gonzalo Mardones Viviani
Graduated from the Catholic University of Chile. Directed degree projects as professor in architectural design at the same university and several other universities. Also he has been working as guest professor and lecturer. Received the first prize in the Architecture Biennale for urban renewal of the South-West center of Santiago.

>>60
H & P Architects
Both Doan Thanh Ha[left] and Tran Ngoc Phuong[right] graduated from Hanoi Architectural University in 2002. They set up and have been operating H&P Architects since 2009. Won some achievements together with H&P Architects; first prize of International Green Architecture Design Competition - FuturArc Prize 2009, second prize of Vietnamese Architects Association Award in 2010, Citation Award of Home for Ecological Living in Asia - FuturArc Prize 2010, second prize of Competition for Bien Hoa High School, Ha Nam, Vietnam in 2011, first prize of the

>>12
Ana Laura Vasconcelos
Was born in Azores in 1979. Graduated from Faculty of Architecture in Oporto in 2004, also studied at the school of Architecture of Barcelona, Technical University of Catalonia, and got academic and professional training in Jordi Bellmunt / Xavi Andreu's architecture studio. Worked for Ainda Arquitectura studio in Oporto. In addition to being currently involved in several private projects, has been working for a company concerning environmental promotion, management and territory planning since 2007.

>>98
Domenico Fiore
Domenico Fiore lives and works in Matera, where he made numerous recovery interventions, especially residences and accommodation facilities. After studying and research activities at the IUAV, has made The manual of recovery of the "Sassi" with Amerigo Restucci and wrote the Inventory of cultural heritage, landscape and environment of the area in Matera. Currently serves as a consultant to the municipality of Matera for preparation of management plan of the UNESCO site "The Sassi and the Park of the Rock Churches of Matera".

Jaap Dawson
Graduated from Cornell University with a Bachelor of Arts in English in 1971. Afterward he earned a Doctor's Degree in Education from Teachers College, Columbia University in 1979. And also received a Ingenieursdiploma(Master of Science) in Architecture from Technische Universiteit Delft in 1988. Currently he delivers a lecture of Architectural Composition in Technische Universiteit Delft and acts as architect, writer, and editor in Delft, the Netherlands since 1988.

Maurizio Scarciglia
Is an Italian architect and urbanist. After several years of professional experience in international offices such as OMA-Rem Koolhaas(Rotterdam), EEA Erick van Egeraat Associated Architects(Rotterdam) and Massimiliano Fuksas(Rome), he founded NAUTA Architecture & Research, with offices in the Netherlands and Italy in 2007. The office engages projects and studies in the fields of architecture, urbanism, art and culture. Since 2008, he has been an external researcher and guest teacher at TU Delft.

TaeCheol Kim
TaeCheol Kim was born in Seoul in 1963. Graduated from HanYang University and received a master's degree from North London in the UK. Started his career at Space Group of Korea from 1985 for which he worked again after coming back from the UK. Also worked for Architecture Environment Group, Hudigm and ran his own office, Leemoon Architecture Studio. Now he is a full-time professor of Dong-A University.

>>122
Isay Weinfeld
Since 1974, he has produced 14 short films, several of which award in Brazil as abroad. Has held a lot of exhibitions in 1990's. Participated in the 25th São Paulo International Biennale and was commissioned as to the exhibition layout for the 26th São Paulo International Biennale. Participated in the 8th Belgrade Triennial of World Architecture in 2006. Served as tenured professor at the Mackenzie Presbyterian University and at the Armando Álvares Penteado Foundation, a Brazilian higher education institution.

>>136
HyoMan Kim
HyoMan Kim is an architect who tries to draw the character of the space and the form of Korean tradition with modern sense. Received a B.A in architecture from DanKook University. Now he is principle of Iroje KHM Architects and current professor of architectural department of SahmYook University. Has received many kinds of architectural awards including World Architecture Community Award and ARCASIA Award.

>>28
Wingårdh Arkitektkontor AB
Is today among the five largest architect groups in Sweden, and among the ten largest in the Nordic region. It has been operating in Goteborg since 1977 and in Stockholm since 1985 by Gert Wingårdh[left]. He was born in 1951 in Skovde, Sweden and received master's degree in architecture from Chalmers University of Technology in Goteborg in 1975. Jonas Edblad[right] was born in 1962. Was employed at Wingårdh in 1990. Received a master's degree in architecture from Chalmers University of Technology, Cothenburg in 1991. Has been a senior lead architect and part of the management team at Wingårdh since 1999.

C3, Issue 2013.9

All Rights Reserved. Authorized translation from the Korean-English language edition published by C3 Publishing Co., Seoul.

© 2013 大连理工大学出版社
著作权合同登记06-2013年第317号

版权所有·侵权必究

图书在版编目(CIP)数据

本土现代化：汉英对照 / 韩国C3出版公社编 ; 李海玲等译. —大连：大连理工大学出版社，2013.12
（C3建筑立场系列丛书）
Vernacular and Modern
ISBN 978-7-5611-8380-9

Ⅰ.①本… Ⅱ.①韩… ②李… Ⅲ.①建筑设计－汉、英 Ⅳ.①TU2

中国版本图书馆CIP数据核字(2013)第300798号

出版发行：大连理工大学出版社
　　　　　（地址：大连市软件园路80号　邮编：116023）
印　　刷：北京雅昌彩色印刷有限公司
幅面尺寸：225mm×300mm
印　　张：12
出版时间：2014年2月第1版
印刷时间：2014年2月第1次印刷
出 版 人：金英伟
统　　筹：房　磊
责任编辑：张昕焱
封面设计：王志峰
责任校对：耿婷婷

书　　号：978-7-5611-8380-9
定　　价：228.00元

发　行：0411-84708842
传　真：0411-84701466
E-mail：dutp@dutp.cn
URL: http://www.dutp.cn